It's another great book from CGP...

This book has everything you'll need to get a solid grounding in Science at KS3 (ages 11-14) — every topic is explained in a clear, straightforward style.

It's ideal if you're working at foundation level — it covers what would have been called Levels 3-6 in the old Curriculum.

CGP — still the best! ☺

Our sole aim here at CGP is to produce the highest quality books — carefully written, immaculately presented and dangerously close to being funny.

Then we work our socks off to get them out to you — at the cheapest possible prices.

Contents

Section 6 — Chemical Changes

Section 7 — The Earth and The Atmosphere

Section 8 — Energy and Matter

Section 9 — Forces and Motion

Section 10 — Waves

Section 11 — Electricity and Magnetism

Section 12 — The Earth and Beyond

Published by CGP

Editors:
David Maliphant, Matteo Orsini Jones, Rachael Rogers, Hayley Thompson.

With thanks to Ian Francis and Jamie Sinclair for the proofreading.

ISBN: 978 1 84146 240 0

www.cgpbooks.co.uk
Clipart from Corel®
Printed by Elanders Ltd, Newcastle upon Tyne.

The Microscope

A microscope is used to look at objects that are <u>too small</u> for you to see normally.
The microscope <u>magnifies</u> the objects (makes them <u>look bigger</u>) so that you can <u>see them</u>.

Learn the Different Parts of a Microscope

This is a microscope:

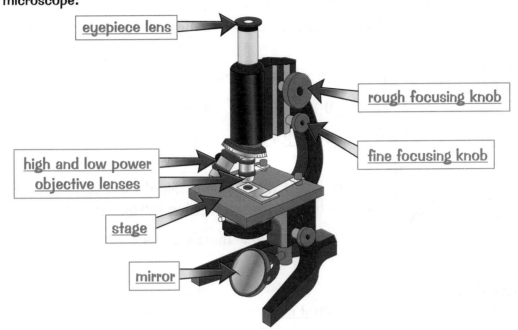

eyepiece lens

rough focusing knob

fine focusing knob

high and low power objective lenses

stage

mirror

Follow These Easy Steps to Using a Microscope

1) Place your microscope near a <u>lamp</u> or a <u>window</u>.
2) Move the mirror so light shines up through the <u>hole</u> in the stage.
3) Get your <u>slide</u> — it should have the object you want to look at <u>stuck to it</u>.
4) Clip your slide to the <u>stage</u>.
5) Select the <u>lowest</u> powered <u>objective lens</u> (the <u>shortest</u> one).

a playground slide

6) <u>Turn</u> the <u>rough focusing knob</u> to move the <u>objective lens</u> down. Stop when the lens is just above the slide.
7) <u>Look down</u> the <u>eyepiece lens</u>.
8) You want to see a <u>clear image</u> of whatever's on the slide. Turn the <u>fine focusing knob</u> until this happens.

a microscope slide

9) If you need to make the image bigger, use a <u>higher powered objective lens</u> (a longer one).
10) Now refocus the microscope (repeat steps 6 to 8).

Microscopes — useful for looking at onions...

Teachers love getting you to look at onions under the microscope. Not a whole one mind, just the <u>slimy skin</u> between the onion layers (yuk). A microscope lets you see all the <u>tiny building blocks</u> (called <u>cells</u>, see next page) that the onion skin is made up of. It's more interesting than it sounds.

Cells

This page is all about what living things are <u>made of</u>. And that includes <u>you</u>.

Living Things are Made of Cells

1) Another word for a <u>living thing</u> is an <u>ORGANISM</u>.

2) <u>All organisms</u> are made up of <u>tiny building blocks</u> called <u>cells</u>.

3) Cells are <u>really small</u>. So you need a <u>microscope</u> to see them (see previous page).

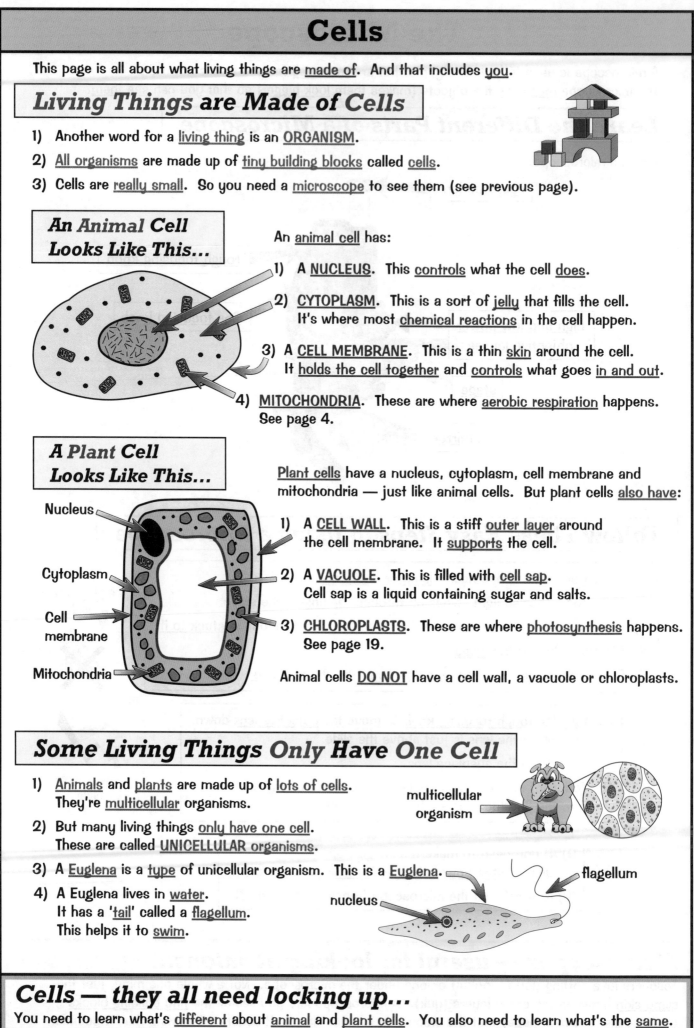

An Animal Cell Looks Like This...

An <u>animal cell</u> has:

1) A <u>NUCLEUS</u>. This <u>controls</u> what the cell <u>does</u>.

2) <u>CYTOPLASM</u>. This is a sort of <u>jelly</u> that fills the cell. It's where most <u>chemical reactions</u> in the cell happen.

3) A <u>CELL MEMBRANE</u>. This is a thin <u>skin</u> around the cell. It <u>holds the cell together</u> and <u>controls</u> what goes <u>in and out</u>.

4) <u>MITOCHONDRIA</u>. These are where <u>aerobic respiration</u> happens. See page 4.

A Plant Cell Looks Like This...

Nucleus

Cytoplasm

Cell membrane

Mitochondria

<u>Plant cells</u> have a nucleus, cytoplasm, cell membrane and mitochondria — just like animal cells. But plant cells <u>also have</u>:

1) A <u>CELL WALL</u>. This is a stiff <u>outer layer</u> around the cell membrane. It <u>supports</u> the cell.

2) A <u>VACUOLE</u>. This is filled with <u>cell sap</u>. Cell sap is a liquid containing sugar and salts.

3) <u>CHLOROPLASTS</u>. These are where <u>photosynthesis</u> happens. See page 19.

Animal cells <u>DO NOT</u> have a cell wall, a vacuole or chloroplasts.

Some Living Things Only Have One Cell

1) <u>Animals</u> and <u>plants</u> are made up of <u>lots of cells</u>. They're <u>multicellular</u> organisms.

2) But many living things <u>only have one cell</u>. These are called <u>UNICELLULAR</u> organisms.

3) A <u>Euglena</u> is a <u>type</u> of unicellular organism. This is a <u>Euglena</u>.

4) A Euglena lives in <u>water</u>. It has a '<u>tail</u>' called a <u>flagellum</u>. This helps it to <u>swim</u>.

multicellular organism

flagellum

nucleus

Cells — they all need locking up...

You need to learn what's <u>different</u> about <u>animal</u> and <u>plant cells</u>. You also need to learn what's the <u>same</u>.

Cell Organisation and Diffusion

I'd like to be a cell — they're always <u>organised</u>. Which is more than can be said for me...

Cells are Organised

In organisms with <u>lots of cells</u>, the cells are <u>organised</u> (sorted) into <u>groups</u>. Here's how:

> 1) A group of <u>similar cells</u> work together to make a <u>tissue</u>.
> 2) A group of <u>different tissues</u> work together to make an <u>organ</u>.
> 3) A <u>group of organs</u> work together to make an <u>organ system</u>.
> 4) A multicellular <u>organism</u> is usually made up of <u>several organ systems</u>.

Here's an example from a <u>plant</u>.

  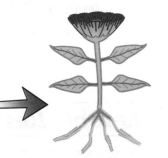

A group of palisade cells make palisade <u>**TISSUE**</u>.

> A palisade cell is just a type of plant cell.

Palisade tissue works with other tissues to make a leaf (an <u>**ORGAN**</u>).

Many leaves and other organs make up the shoot system, an <u>**ORGAN SYSTEM**</u>.

Different organ systems make up a plant — an <u>**ORGANISM**</u>.

Stuff Moves Into and Out of Cells by Diffusion

1) Substances <u>move into</u> or <u>out of cells</u> by a process called <u>diffusion</u>.

2) Diffusion is where stuff moves from where there's <u>lots of it</u> to where there's <u>less of it</u>. Just like <u>glucose</u> in this diagram...

> There's more on diffusion on pages 33 and 73.

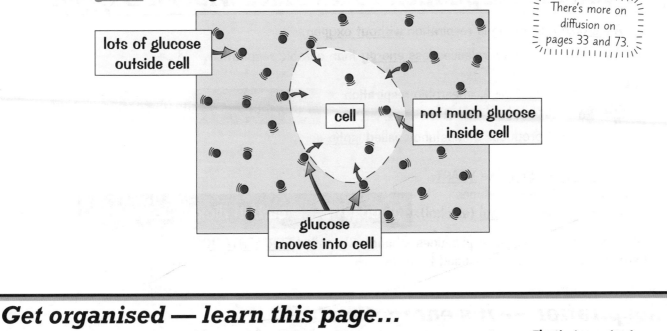

lots of glucose outside cell

cell

not much glucose inside cell

glucose moves into cell

Get organised — learn this page...

Remember: cells ⟹ tissues ⟹ organs ⟹ organ systems ⟹ organisms. That's <u>important</u>. The example shown at the top of the page is for plants, but <u>animal cells</u> are organised in the <u>same way</u>.

Respiration

Respiration is important. Without it you <u>wouldn't have energy</u> to do stuff, like learning KS3 science...

Respiration is a Chemical Reaction

1) In a chemical reaction one or more <u>'old' substances</u> get <u>changed</u> into <u>new ones</u>.
   The old substances are called <u>reactants</u>. The new substances are called <u>products</u>. See page 45.

2) <u>Respiration</u> is a <u>chemical reaction</u>. It happens in <u>every cell</u>.

3) Respiration changes <u>glucose</u> (a sugar) into new substances. This <u>releases energy</u>.

4) The <u>energy</u> released by respiration is used for just about <u>everything</u>. For example:

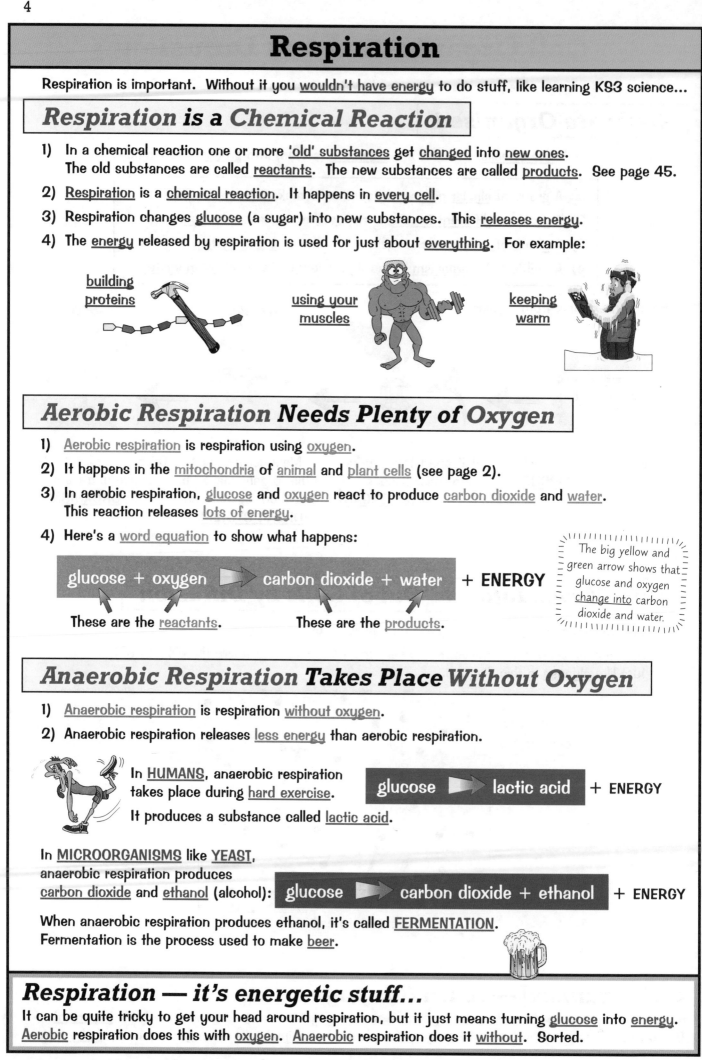

<u>building proteins</u>

<u>using your muscles</u>

<u>keeping warm</u>

Aerobic Respiration Needs Plenty of Oxygen

1) <u>Aerobic respiration</u> is respiration using <u>oxygen</u>.

2) It happens in the <u>mitochondria</u> of <u>animal</u> and <u>plant cells</u> (see page 2).

3) In aerobic respiration, <u>glucose</u> and <u>oxygen</u> react to produce <u>carbon dioxide</u> and <u>water</u>.
   This reaction releases <u>lots of energy</u>.

4) Here's a <u>word equation</u> to show what happens:

glucose + oxygen ➜ carbon dioxide + water **+ ENERGY**

These are the <u>reactants</u>.   These are the <u>products</u>.

The big yellow and green arrow shows that glucose and oxygen <u>change into</u> carbon dioxide and water.

Anaerobic Respiration Takes Place Without Oxygen

1) <u>Anaerobic respiration</u> is respiration <u>without oxygen</u>.

2) Anaerobic respiration releases <u>less energy</u> than aerobic respiration.

In <u>HUMANS</u>, anaerobic respiration takes place during <u>hard exercise</u>.

It produces a substance called <u>lactic acid</u>.

glucose ➜ lactic acid **+ ENERGY**

In <u>MICROORGANISMS</u> like <u>YEAST</u>, anaerobic respiration produces <u>carbon dioxide</u> and <u>ethanol</u> (alcohol):

glucose ➜ carbon dioxide + ethanol **+ ENERGY**

When anaerobic respiration produces ethanol, it's called <u>FERMENTATION</u>.
Fermentation is the process used to make <u>beer</u>.

Respiration — it's energetic stuff...

It can be quite tricky to get your head around respiration, but it just means turning <u>glucose</u> into <u>energy</u>.
<u>Aerobic</u> respiration does this with <u>oxygen</u>. <u>Anaerobic</u> respiration does it <u>without</u>. Sorted.

Section Summary

Welcome to your very first Section Summary. It's full of questions that will help you find out what you actually know — and, more importantly, what you don't. Here's what you have to do...

1) Go through the whole lot of these Section Summary questions and try to answer them.
2) Look up the answers to any you can't do and try to really learn them
   (hint: the answers are all somewhere in Section 1).
3) Try all the questions again to see if you can answer more than you could before.
4) Keep going till you get them all right.

Better get started then...

1) What part of a microscope do you clip your slide onto?
2) What do the focusing knobs on a microscope do?
3) How do you make the image you see through a microscope look bigger?
4) What is an organism?
5) What piece of equipment would you use to look at a cell?
6) Name four parts of a cell that both plant cells and animal cells have. Say what they all do.
7) Name three parts of a cell that only plant cells have.
8) Name a unicellular organism. Draw a picture to show what it looks like.
9) Explain the meaning of: a) tissue   b) organ c) organ system. Give an example of each.
10) What is diffusion?
11) Give an example of a substance that moves into or out of cells by diffusion.
12) What's the name of the chemical reaction that goes on in every cell and releases energy?
13) What is the energy released by this reaction used for? Give three examples.
14) What is aerobic respiration? Where does it take place in plant and animal cells?
15) Write down the word equation for aerobic respiration.
16) What is anaerobic respiration?
17) What substance is produced by anaerobic respiration in humans?
18) What are the products of anaerobic respiration in yeast?
19) What is fermentation? What can fermentation be used to make?

Nutrition

Nutrition is <u>what you eat</u> — and what you eat is really <u>important</u> for your <u>health</u>.

A Balanced Diet Contains All These Things

A <u>balanced diet</u> will have the right amount of the <u>five nutrients</u> below:

Nutrient	What it's found in	What it's needed for
Carbohydrates	Bread, potatoes, cereals	You need <u>lots</u> of carbohydrate if you're <u>active</u> or <u>growing</u>. Energy
Proteins	Meat, eggs, fish	You need proteins to <u>grow</u> and to <u>repair</u> damage.
Lipids (fats and oils)	Butter, cooking oil, cream	You use lipids for energy if your body <u>runs out</u> of <u>carbohydrates</u>. Energy
Vitamins e.g. Vitamin A, Vitamin C	Vegetables, fruit, cereals	Vitamins keep many <u>important</u> <u>processes</u> happening in your body.
Minerals e.g. calcium, iron	For example: • <u>calcium</u> is found in milk, • <u>iron</u> is found in meat.	<u>Minerals</u> are needed for lots of things. For example: • <u>calcium</u> is needed for strong <u>bones</u> and <u>teeth</u>, • <u>iron</u> is needed for healthy <u>blood</u>.

A <u>balanced diet</u> will also have enough <u>fibre</u> and <u>water</u>:

	What it's found in	What it's needed for
Fibre	Vegetables, fruit, cereals	Fibre helps food <u>move</u> through your <u>digestive system</u>.
Water	Drinks, watery foods like soup	All the <u>chemical reactions</u> in your body happen in water.

You'll go all wobbly if your diet isn't balanced...

Five types of <u>nutrient</u>, what you <u>find them in</u> and what they're <u>for</u>. And then there's <u>fibre</u> and <u>water</u> too. <u>Learn</u> it all with the <u>only method</u> that works — <u>covering the page</u> and <u>jotting it down</u>.

More on Nutrition

Your body needs energy <u>all the time</u>. You get energy from <u>carbohydrates</u> and <u>fats</u> in your <u>diet</u>.

Different People Have Different Energy Needs

1) The <u>heavier</u> you are, the <u>more energy</u> you will need.

2) Also, the <u>more active</u> you are, the <u>more energy</u> you will need.

You Can Work Out Your Daily Basic Energy Requirement

1) Your daily <u>basic energy requirement (BER)</u> is the energy you need every day <u>just to stay alive</u>.

2) You calculate BER like this: ➡ Daily BER (kJ/day) = 5.4 × 24 hours × body mass (kg)

A kJ is a unit of energy.

EXAMPLE: Work out the daily BER for a <u>60 kg</u> person.
ANSWER: Daily BER = 5.4 × 24 × 60 = <u>7776 kJ/day</u>.

You Need Extra Energy for Your Activities

For example:

<u>Walking</u> for half an hour uses <u>400 kJ</u> of energy.

<u>Running</u> for half an hour uses <u>1500 kJ</u> of energy.

The <u>total amount of energy</u> you need in a day = <u>daily BER</u> + <u>extra energy</u> for <u>activities</u>.

An Unbalanced Diet Can Cause Health Problems

Obesity

1) If you <u>take in more energy</u> than you <u>use up</u>, you will <u>put on weight</u>.

2) Over time you could become <u>obese</u> (very overweight).

3) Obesity can lead to <u>health problems</u> such as <u>heart disease</u>.

Starvation

1) Some people don't get <u>enough food to eat</u> — this is <u>starvation</u>.

2) Starvation can cause <u>slow growth</u> in children and <u>irregular periods</u> in women.

Deficiency Diseases

1) Some people don't get enough <u>vitamins or minerals</u> — this can cause <u>deficiency diseases</u>.

2) For example, not getting enough <u>vitamin C</u> can cause <u>scurvy</u>.
This is a deficiency disease that causes problems with the <u>skin</u> and <u>gums</u>.

Summon up the energy to learn this page...

You should be able to <u>work out</u> how much <u>energy</u> a person needs in a day. Also, make sure you understand the <u>health problems</u> that can be caused by eating <u>too much food</u> or <u>too little food</u>.

Digestion

Digestion is all about <u>breaking food down</u> so we can use the <u>nutrients</u> it contains.
But it's not easy — lots of different <u>organs</u> have to <u>work together</u> to get the job done.

There are Two Steps to Digestion

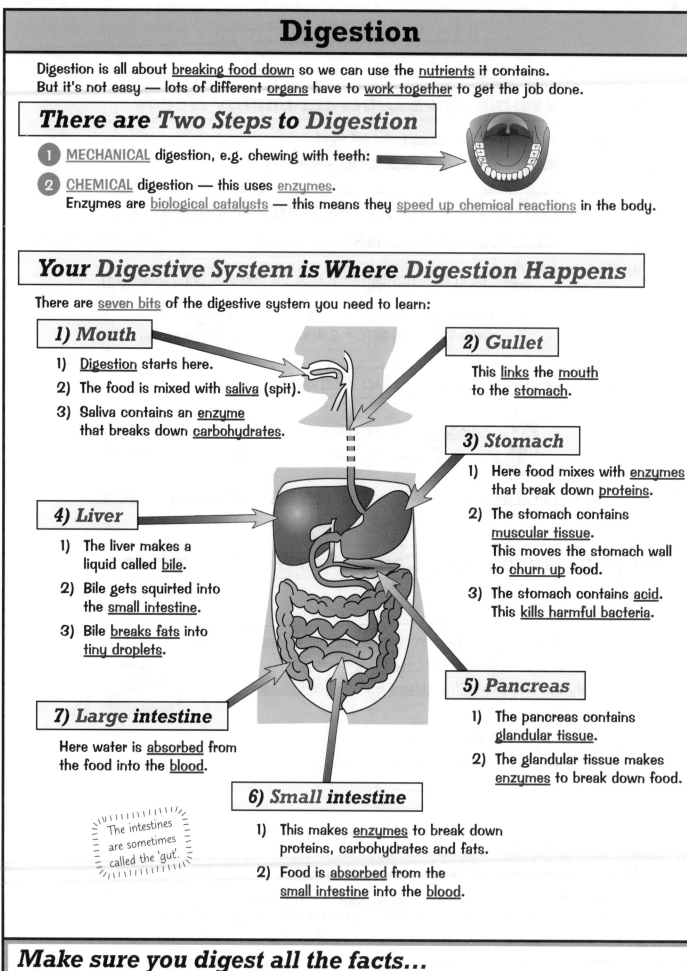

1 <u>MECHANICAL</u> digestion, e.g. chewing with teeth:

2 <u>CHEMICAL</u> digestion — this uses <u>enzymes</u>.
Enzymes are <u>biological catalysts</u> — this means they <u>speed up chemical reactions</u> in the body.

Your Digestive System is Where Digestion Happens

There are <u>seven bits</u> of the digestive system you need to learn:

1) Mouth

1) <u>Digestion</u> starts here.

2) The food is mixed with <u>saliva</u> (spit).

3) Saliva contains an <u>enzyme</u>
that breaks down <u>carbohydrates</u>.

2) Gullet

This <u>links</u> the <u>mouth</u>
to the <u>stomach</u>.

3) Stomach

1) Here food mixes with <u>enzymes</u>
that break down <u>proteins</u>.

2) The stomach contains
<u>muscular tissue</u>.
This moves the stomach wall
to <u>churn up</u> food.

3) The stomach contains <u>acid</u>.
This <u>kills harmful bacteria</u>.

4) Liver

1) The liver makes a
liquid called <u>bile</u>.

2) Bile gets squirted into
the <u>small intestine</u>.

3) Bile <u>breaks fats</u> into
<u>tiny droplets</u>.

5) Pancreas

1) The pancreas contains
<u>glandular tissue</u>.

2) The glandular tissue makes
<u>enzymes</u> to break down food.

7) Large intestine

Here water is <u>absorbed</u> from
the food into the <u>blood</u>.

The intestines
are sometimes
called the 'gut'.

6) Small intestine

1) This makes <u>enzymes</u> to break down
proteins, carbohydrates and fats.

2) Food is <u>absorbed</u> from the
<u>small intestine</u> into the <u>blood</u>.

Make sure you digest all the facts...

Start by learning <u>the headings</u> — including each of the <u>seven bits</u> of the digestive system.
Then <u>learn all the details</u> that go with them. Write down as much as you can remember, then
check the page to see what you've missed. Keep doing this until you can <u>get it all</u> right.

More on Digestion

Well <u>would you believe it</u>? There's more to learn about digestion.

Food Molecules *Get Absorbed* in the *Small Intestine*

1) <u>Big</u> food molecules <u>can't</u> fit through the <u>small intestine wall</u>.

2) So enzymes <u>break up</u> the <u>big molecules</u> into <u>smaller molecules</u>.

3) The small molecules <u>pass through</u> the small intestine wall into the <u>blood</u>.

4) They then travel round the body in the <u>blood</u> to <u>body cells</u>, where they are used.

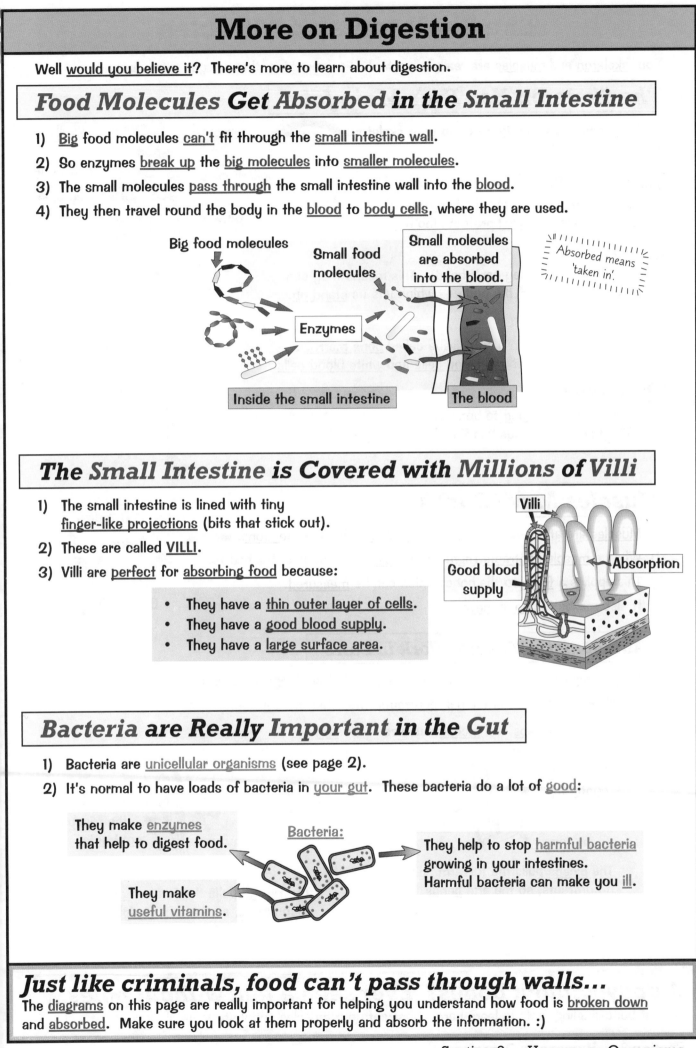

Big food molecules

Small food molecules

Small molecules are absorbed into the blood.

Absorbed means 'taken in'.

Enzymes

Inside the small intestine

The blood

The *Small Intestine* is *Covered* with *Millions* of *Villi*

1) The small intestine is lined with tiny <u>finger-like projections</u> (bits that stick out).

2) These are called <u>VILLI</u>.

3) Villi are <u>perfect</u> for <u>absorbing food</u> because:

- They have a <u>thin outer layer of cells</u>.
- They have a <u>good blood supply</u>.
- They have a <u>large surface area</u>.

Villi

Good blood supply

Absorption

Bacteria are *Really Important* in the *Gut*

1) Bacteria are <u>unicellular organisms</u> (see page 2).

2) It's normal to have loads of bacteria in <u>your gut</u>. These bacteria do a lot of <u>good</u>:

They make <u>enzymes</u> that help to digest food.

Bacteria:

They help to stop <u>harmful bacteria</u> growing in your intestines. Harmful bacteria can make you <u>ill</u>.

They make <u>useful vitamins</u>.

Just like criminals, food can't pass through walls...

The <u>diagrams</u> on this page are really important for helping you understand how food is <u>broken down</u> and <u>absorbed</u>. Make sure you look at them properly and absorb the information. :)

The Skeleton and Muscles

Your <u>skeleton</u> and <u>muscles</u> are really important. For one thing, they let you <u>move around</u>.

The Skeleton Has Four Main Jobs

All the <u>bones</u> in your body make up your <u>skeleton</u>.

The skeleton's jobs are:

1 PROTECTION:

Bone is <u>tough</u>, so it can <u>protect organs</u>.
For example, the <u>skull</u> protects the <u>brain</u>.

2 SUPPORT:

Bones are <u>rigid</u> (they can't bend). This means they can support the rest of the body — which lets us <u>stand up</u>.

3 MAKING BLOOD CELLS:

Many bones contain a soft tissue called <u>bone marrow</u>.
Bone marrow makes <u>red blood cells</u> and <u>white blood cells</u>.

4 MOVEMENT:

<u>Muscles</u> are <u>attached</u> to bones.
The action of muscles lets the skeleton <u>move</u>.

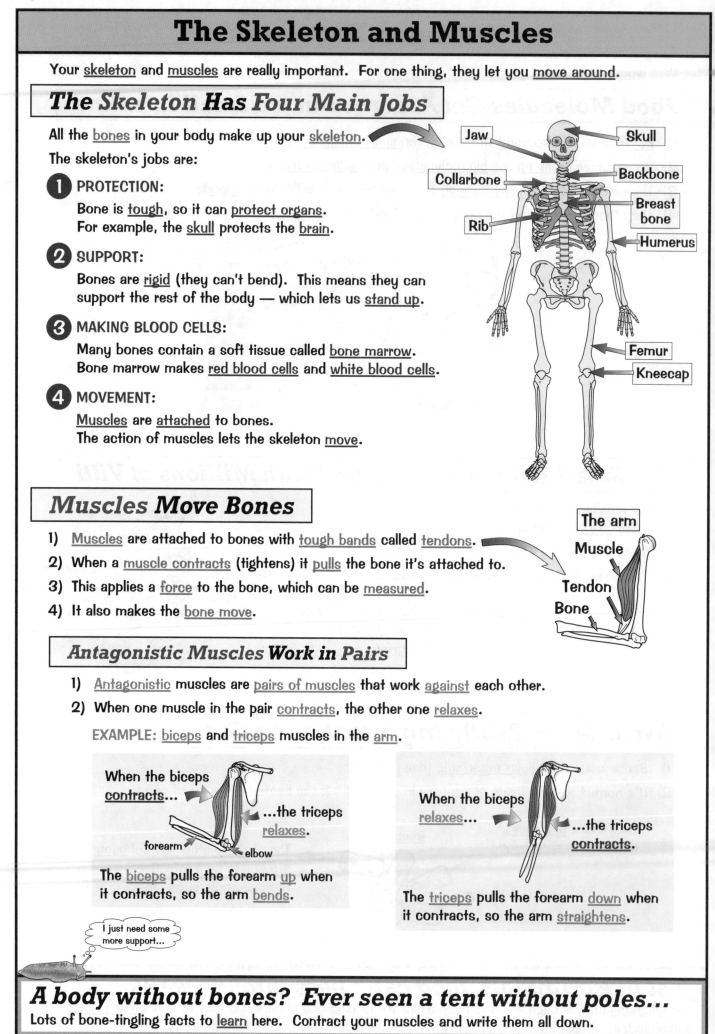

Jaw | Skull | Collarbone | Backbone | Breast bone | Rib | Humerus | Femur | Kneecap

Muscles Move Bones

1) <u>Muscles</u> are attached to bones with <u>tough bands</u> called <u>tendons</u>.
2) When a <u>muscle contracts</u> (tightens) it <u>pulls</u> the bone it's attached to.
3) This applies a <u>force</u> to the bone, which can be <u>measured</u>.
4) It also makes the <u>bone move</u>.

The arm
Muscle
Tendon
Bone

Antagonistic Muscles Work in Pairs

1) <u>Antagonistic</u> muscles are <u>pairs of muscles</u> that work <u>against</u> each other.
2) When one muscle in the pair <u>contracts</u>, the other one <u>relaxes</u>.

EXAMPLE: <u>biceps</u> and <u>triceps</u> muscles in the <u>arm</u>.

When the biceps <u>contracts</u>... ...the triceps <u>relaxes</u>.
forearm   elbow

The <u>biceps</u> pulls the forearm <u>up</u> when it contracts, so the arm <u>bends</u>.

When the biceps <u>relaxes</u>... ...the triceps <u>contracts</u>.

The <u>triceps</u> pulls the forearm <u>down</u> when it contracts, so the arm <u>straightens</u>.

I just need some more support...

A body without bones? Ever seen a tent without poles...

Lots of bone-tingling facts to <u>learn</u> here. Contract your muscles and write them all down.

Gas Exchange

You need to get <u>oxygen</u> from the air into your blood. You also need to get rid of the <u>carbon dioxide</u> that's in your blood. That's why you need your <u>gas exchange system</u>.

Learn *These* Structures *in the Gas Exchange System*

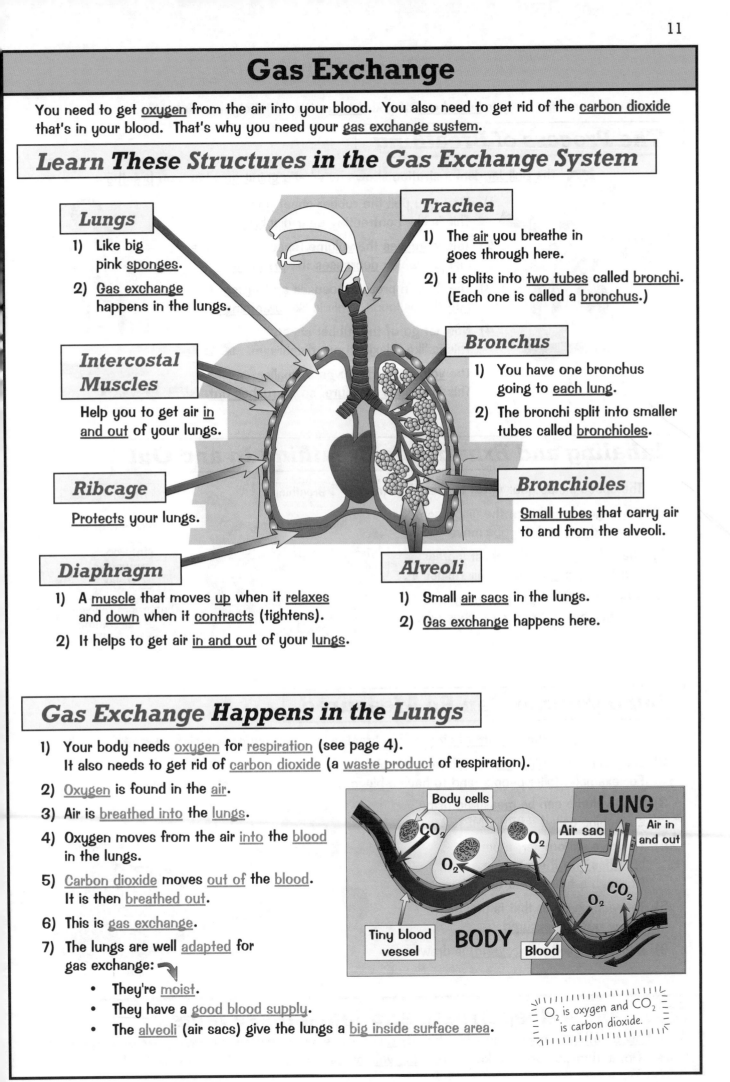

Lungs

1) Like big pink <u>sponges</u>.

2) <u>Gas exchange</u> happens in the lungs.

Trachea

1) The <u>air</u> you breathe in goes through here.

2) It splits into <u>two tubes</u> called <u>bronchi</u>. (Each one is called a <u>bronchus</u>.)

Intercostal Muscles

Help you to get air <u>in</u> <u>and out</u> of your lungs.

Bronchus

1) You have one bronchus going to <u>each lung</u>.

2) The bronchi split into smaller tubes called <u>bronchioles</u>.

Ribcage

<u>Protects</u> your lungs.

Bronchioles

<u>Small tubes</u> that carry air to and from the alveoli.

Diaphragm

1) A <u>muscle</u> that moves <u>up</u> when it <u>relaxes</u> and <u>down</u> when it <u>contracts</u> (tightens).

2) It helps to get air <u>in and out</u> of your <u>lungs</u>.

Alveoli

1) Small <u>air sacs</u> in the lungs.

2) <u>Gas exchange</u> happens here.

Gas Exchange *Happens in the Lungs*

1) Your body needs <u>oxygen</u> for <u>respiration</u> (see page 4). It also needs to get rid of <u>carbon dioxide</u> (a <u>waste product</u> of respiration).

2) <u>Oxygen</u> is found in the <u>air</u>.

3) Air is <u>breathed into</u> the <u>lungs</u>.

4) Oxygen moves from the air <u>into</u> the <u>blood</u> in the lungs.

5) <u>Carbon dioxide</u> moves <u>out of</u> the <u>blood</u>. It is then <u>breathed out</u>.

6) This is <u>gas exchange</u>.

7) The lungs are well <u>adapted</u> for gas exchange:

 • They're <u>moist</u>.
 • They have a <u>good blood supply</u>.
 • The <u>alveoli</u> (air sacs) give the lungs a <u>big inside surface area</u>.

O_2 is oxygen and CO_2 is carbon dioxide.

Breathing

Breathing is how the air gets <u>in and out</u> of your <u>lungs</u>. It's definitely a useful skill.

The Process of Breathing

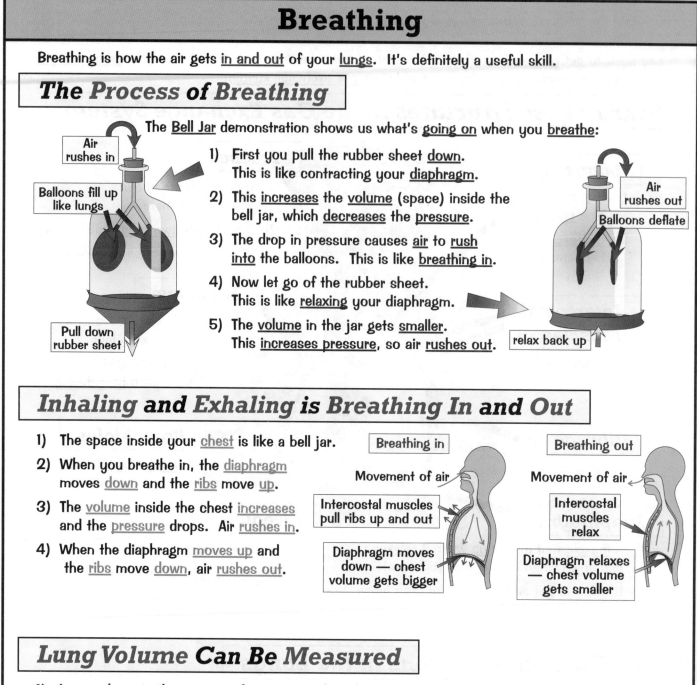

The <u>Bell Jar</u> demonstration shows us what's <u>going on</u> when you <u>breathe</u>:

1) First you pull the rubber sheet <u>down</u>. This is like contracting your <u>diaphragm</u>.

2) This <u>increases</u> the <u>volume</u> (space) inside the bell jar, which <u>decreases</u> the <u>pressure</u>.

3) The drop in pressure causes <u>air</u> to <u>rush</u> <u>into</u> the balloons. This is like <u>breathing in</u>.

4) Now let go of the rubber sheet. This is like <u>relaxing</u> your diaphragm.

5) The <u>volume</u> in the jar gets <u>smaller</u>. This <u>increases pressure</u>, so air <u>rushes out</u>.

Air rushes in

Balloons fill up like lungs

Pull down rubber sheet

Air rushes out

Balloons deflate

relax back up

Inhaling and Exhaling is Breathing In and Out

1) The space inside your <u>chest</u> is like a bell jar.

2) When you breathe in, the <u>diaphragm</u> moves <u>down</u> and the <u>ribs</u> move <u>up</u>.

3) The <u>volume</u> inside the chest <u>increases</u> and the <u>pressure</u> drops. Air <u>rushes in</u>.

4) When the diaphragm <u>moves up</u> and the <u>ribs</u> move <u>down</u>, air <u>rushes out</u>.

Breathing in

Movement of air

Intercostal muscles pull ribs up and out

Diaphragm moves down — chest volume gets bigger

Breathing out

Movement of air

Intercostal muscles relax

Diaphragm relaxes — chest volume gets smaller

Lung Volume Can Be Measured

1) <u>Lung volume</u> is the <u>amount of air</u> you can breathe into your lungs in a single breath.

2) Lung volume is <u>different for different people</u>. For example, <u>taller</u> people tend to have a <u>bigger</u> lung volume than <u>shorter</u> people.

3) Lung volume can be <u>measured</u> using a machine called a <u>SPIROMETER</u>.

4) To use a spirometer, a person <u>breathes into the machine</u> (through a tube) for a few minutes.

5) The volume of air that is breathed in and out is <u>measured</u>.

6) A graph (called a <u>spirogram</u>) is drawn.

spirometer

spirogram

Now take a deep breath and learn these facts...

Learn <u>how breathing works</u> — use that <u>bell jar demo</u> to help you understand what goes on in your actual lungs. Oh, and make sure you know how <u>lung volume</u> can be measured too. It's all good fun.

Exercise, Asthma and Smoking

Exercise, asthma and smoking can all affect your <u>gas exchange system</u> (see p. 11) and your <u>breathing</u>.

Exercise

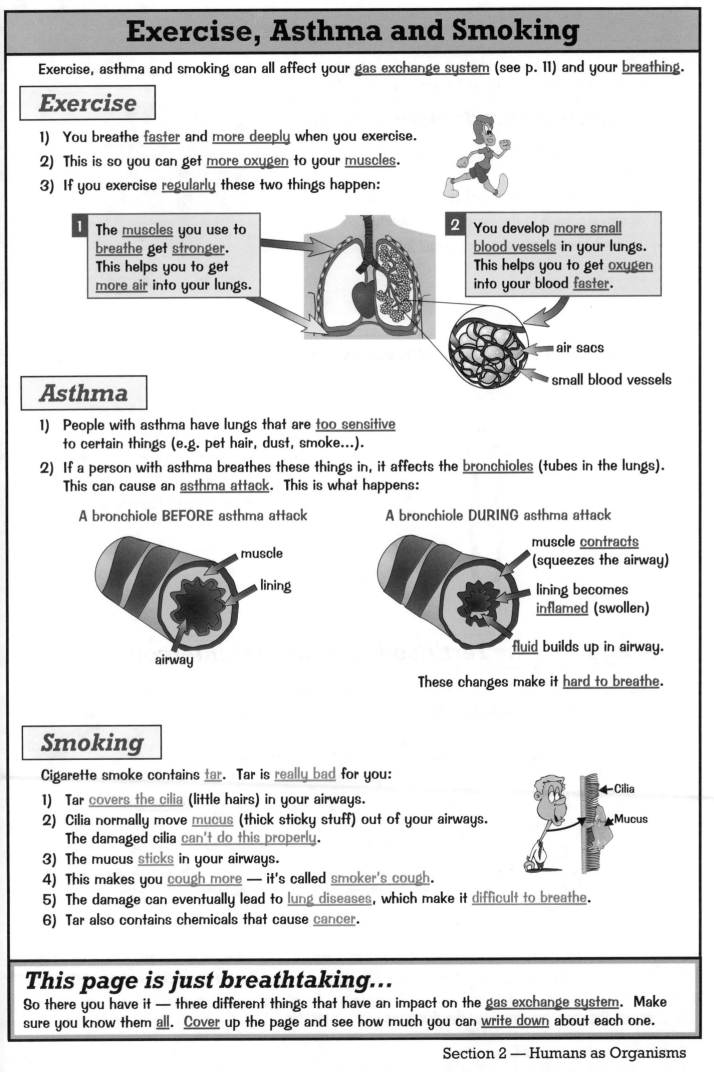

1) You breathe <u>faster</u> and <u>more deeply</u> when you exercise.

2) This is so you can get <u>more oxygen</u> to your <u>muscles</u>.

3) If you exercise <u>regularly</u> these two things happen:

1 The <u>muscles</u> you use to <u>breathe</u> get <u>stronger</u>. This helps you to get <u>more air</u> into your lungs.

2 You develop <u>more small blood vessels</u> in your lungs. This helps you to get <u>oxygen</u> into your blood <u>faster</u>.

air sacs

small blood vessels

Asthma

1) People with asthma have lungs that are <u>too sensitive</u> to certain things (e.g. pet hair, dust, smoke...).

2) If a person with asthma breathes these things in, it affects the <u>bronchioles</u> (tubes in the lungs). This can cause an <u>asthma attack</u>. This is what happens:

A bronchiole BEFORE asthma attack

muscle

lining

airway

A bronchiole DURING asthma attack

muscle <u>contracts</u> (squeezes the airway)

lining becomes <u>inflamed</u> (swollen)

<u>fluid</u> builds up in airway.

These changes make it <u>hard to breathe</u>.

Smoking

Cigarette smoke contains <u>tar</u>. Tar is <u>really bad</u> for you:

1) Tar <u>covers the cilia</u> (little hairs) in your airways.

2) Cilia normally move <u>mucus</u> (thick sticky stuff) out of your airways. The damaged cilia <u>can't do this properly</u>.

3) The mucus <u>sticks</u> in your airways.

4) This makes you <u>cough more</u> — it's called <u>smoker's cough</u>.

5) The damage can eventually lead to <u>lung diseases</u>, which make it <u>difficult to breathe</u>.

6) Tar also contains chemicals that cause <u>cancer</u>.

Cilia

Mucus

This page is just breathtaking...

So there you have it — three different things that have an impact on the <u>gas exchange system</u>. Make sure you know them <u>all</u>. <u>Cover</u> up the page and see how much you can <u>write down</u> about each one.

Human Reproduction

Like all <u>mammals</u>, we have different <u>boy bits</u> and <u>girl bits</u> that allow us to <u>reproduce</u> (make babies). Humans reproduce using sexual intercourse. No giggling now.

The Male Reproductive System

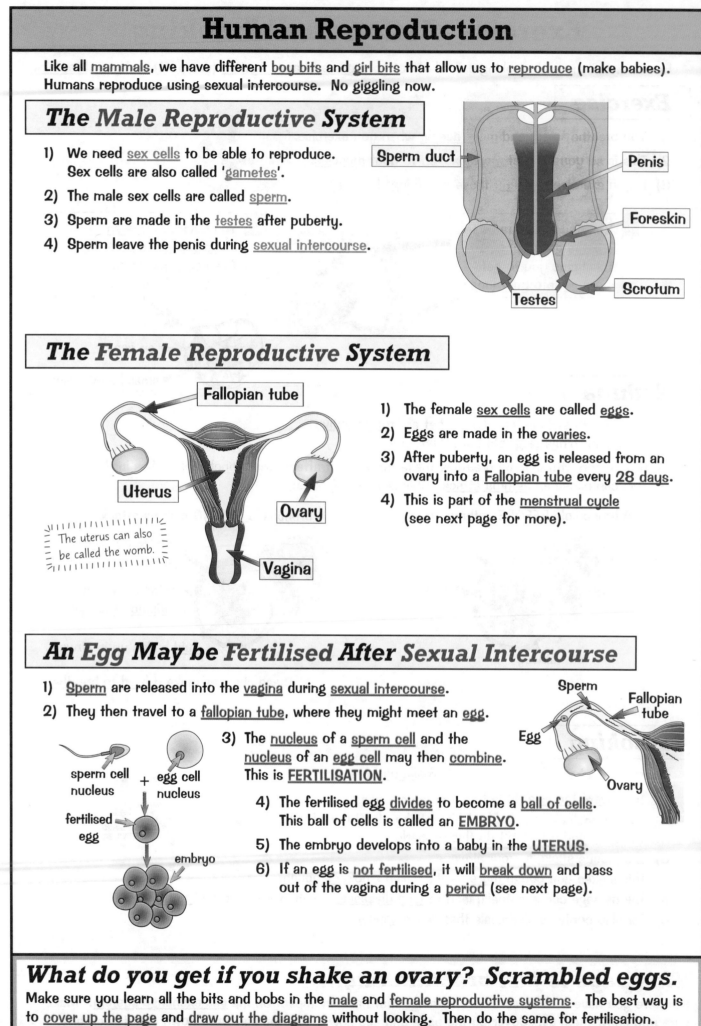

1) We need <u>sex cells</u> to be able to reproduce. Sex cells are also called '<u>gametes</u>'.

2) The male sex cells are called <u>sperm</u>.

3) Sperm are made in the <u>testes</u> after puberty.

4) Sperm leave the penis during <u>sexual intercourse</u>.

Sperm duct

Penis

Foreskin

Scrotum

Testes

The Female Reproductive System

Fallopian tube

Uterus

Ovary

Vagina

The uterus can also be called the womb.

1) The female <u>sex cells</u> are called <u>eggs</u>.

2) Eggs are made in the <u>ovaries</u>.

3) After puberty, an egg is released from an ovary into a <u>Fallopian tube</u> every <u>28 days</u>.

4) This is part of the <u>menstrual cycle</u> (see next page for more).

An Egg May be Fertilised After Sexual Intercourse

1) <u>Sperm</u> are released into the <u>vagina</u> during <u>sexual intercourse</u>.

2) They then travel to a <u>fallopian tube</u>, where they might meet an <u>egg</u>.

sperm cell nucleus + egg cell nucleus

fertilised egg

embryo

Sperm

Fallopian tube

Egg

Ovary

3) The <u>nucleus</u> of a <u>sperm cell</u> and the <u>nucleus</u> of an <u>egg cell</u> may then <u>combine</u>. This is <u>FERTILISATION</u>.

4) The fertilised egg <u>divides</u> to become a <u>ball of cells</u>. This ball of cells is called an <u>EMBRYO</u>.

5) The embryo develops into a baby in the <u>UTERUS</u>.

6) If an egg is <u>not fertilised</u>, it will <u>break down</u> and pass out of the vagina during a <u>period</u> (see next page).

What do you get if you shake an ovary? Scrambled eggs.

Make sure you learn all the bits and bobs in the <u>male</u> and <u>female reproductive systems</u>. The best way is to <u>cover up the page</u> and <u>draw out the diagrams</u> without looking. Then do the same for fertilisation.

The Menstrual Cycle

From the age of puberty, <u>women</u> undergo a <u>monthly</u> sequence of events called the <u>menstrual cycle</u>.

The Menstrual Cycle Takes 28 Days

1) In the menstrual cycle the body <u>prepares</u> the uterus in case it receives a <u>fertilised egg</u>.

2) The diagram below shows the <u>four main stages</u> of the menstrual cycle:

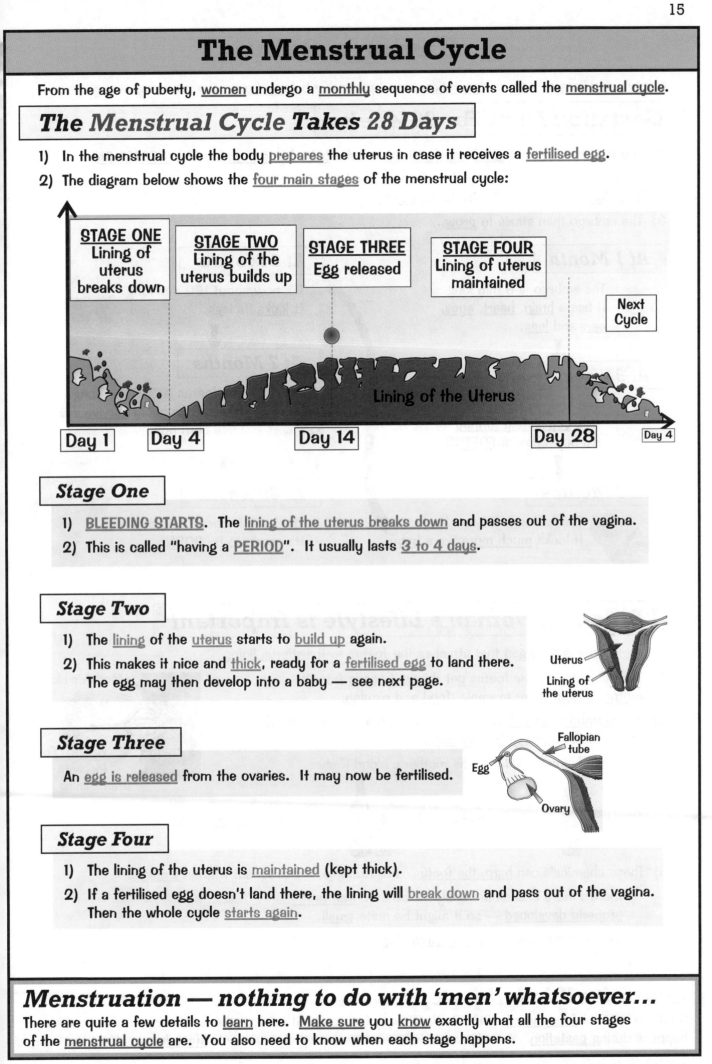

STAGE ONE
Lining of uterus breaks down

STAGE TWO
Lining of the uterus builds up

STAGE THREE
Egg released

STAGE FOUR
Lining of uterus maintained

Next Cycle

Lining of the Uterus

Day 1   Day 4   Day 14   Day 28   Day 4

Stage One

1) <u>BLEEDING STARTS</u>. The <u>lining of the uterus breaks down</u> and passes out of the vagina.

2) This is called "having a <u>PERIOD</u>". It usually lasts <u>3 to 4 days</u>.

Stage Two

1) The <u>lining</u> of the <u>uterus</u> starts to <u>build up</u> again.

2) This makes it nice and <u>thick</u>, ready for a <u>fertilised egg</u> to land there. The egg may then develop into a baby — see next page.

Uterus

Lining of the uterus

Stage Three

An <u>egg is released</u> from the ovaries. It may now be fertilised.

Fallopian tube

Egg

Ovary

Stage Four

1) The lining of the uterus is <u>maintained</u> (kept thick).

2) If a fertilised egg doesn't land there, the lining will <u>break down</u> and pass out of the vagina. Then the whole cycle <u>starts again</u>.

Menstruation — nothing to do with 'men' whatsoever...

There are quite a few details to <u>learn</u> here. <u>Make sure</u> you <u>know</u> exactly what all the four stages of the <u>menstrual cycle</u> are. You also need to know when each stage happens.

Having a Baby

A sperm fertilises an egg, the gestation period passes, a baby is born. Sounds easy enough.

Gestation Lasts For 39 Weeks

1) The time between the egg being fertilised and the baby being born is called GESTATION.

2) Once a fertilised egg has developed into an embryo (see page 14), it implants (sticks itself) into the uterus lining.

3) The embryo then starts to grow...

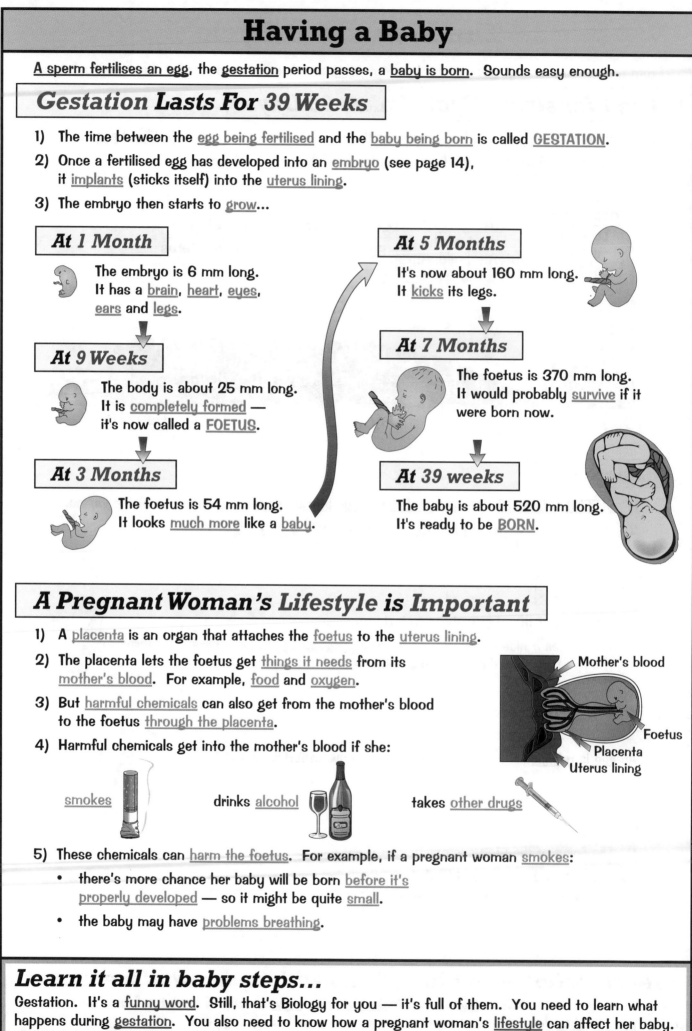

At 1 Month

The embryo is 6 mm long. It has a brain, heart, eyes, ears and legs.

At 9 Weeks

The body is about 25 mm long. It is completely formed — it's now called a FOETUS.

At 3 Months

The foetus is 54 mm long. It looks much more like a baby.

At 5 Months

It's now about 160 mm long. It kicks its legs.

At 7 Months

The foetus is 370 mm long. It would probably survive if it were born now.

At 39 weeks

The baby is about 520 mm long. It's ready to be BORN.

A Pregnant Woman's Lifestyle is Important

1) A placenta is an organ that attaches the foetus to the uterus lining.

2) The placenta lets the foetus get things it needs from its mother's blood. For example, food and oxygen.

3) But harmful chemicals can also get from the mother's blood to the foetus through the placenta.

4) Harmful chemicals get into the mother's blood if she:

smokes   drinks alcohol   takes other drugs

Mother's blood
Foetus
Placenta
Uterus lining

5) These chemicals can harm the foetus. For example, if a pregnant woman smokes:

- there's more chance her baby will be born before it's properly developed — so it might be quite small.

- the baby may have problems breathing.

Learn it all in baby steps...

Gestation. It's a funny word. Still, that's Biology for you — it's full of them. You need to learn what happens during gestation. You also need to know how a pregnant woman's lifestyle can affect her baby.

Health and Drugs

Good health is when both your <u>body</u> and your <u>mind</u> are <u>fine and dandy</u>.
Recreational drugs can have serious, <u>bad effects</u> on your health.

Health is More Than Just Not Being Ill

1) Good health means having: ▶
   - A <u>healthy body</u> that's <u>all working properly</u> with <u>no diseases</u>.
   - A <u>healthy mind</u> so you can cope with the <u>ups and downs</u> of life.

2) Taking <u>drugs</u> can affect your health.

Drugs

1) A drug is anything that <u>affects the way</u> the body works.
   For example, a drug may increase heart rate.

2) Drugs can affect <u>LIFE PROCESSES</u>.
   For example, drugs that affect the <u>brain</u> are likely to
   affect <u>movement</u> and <u>sensitivity</u>.

3) <u>RECREATIONAL DRUGS</u> are drugs used for fun.
   They can be <u>legal</u> (like alcohol) or <u>illegal</u> (like ecstasy).

> **7 Life Processes**
> Movement — moving parts of the body.
> Reproduction — producing offspring.
> Sensitivity — responding and reacting.
> Nutrition — getting food to stay alive.
> Excretion — getting rid of waste.
> Respiration — turning food into energy.
> Growth — getting to adult size.

Solvents

1) Solvents are found in things like <u>paints</u> and <u>glues</u>.

2) Sniffing solvents can make you <u>see</u> and <u>hear</u> things that are <u>not really there</u>.
   <u>Misusing</u> solvents like this can affect your <u>behaviour</u>.

3) Solvents also <u>damage</u> the <u>lungs</u>, <u>brain</u> and <u>kidneys</u>.

Alcohol

1) Alcohol is found in <u>beers</u>, <u>wines</u> and <u>spirits</u>.

2) It <u>decreases brain activity</u>. This means you <u>react</u> to things <u>more slowly</u>.

3) It can damage the <u>brain</u> and <u>liver</u>.

4) It <u>impairs judgement</u> — so you might end up doing silly things.

Illegal Drugs

1) There are many <u>illegal</u> recreational drugs. For example, <u>ecstasy</u>, <u>heroin</u> and <u>LSD</u>.

2) Many illegal drugs are <u>very addictive</u>.
   This means the user feels like they <u>NEED</u> to have them.

3) They affect <u>behaviour</u>.

4) They can have very <u>bad effects</u> on a person's <u>body</u>. For example, <u>ecstasy</u> can
   lead to <u>dehydration</u> (not enough water in the body), which can cause <u>DEATH</u>.

Drugs aren't harmless fun — they're a slippery slope...

Make sure you know what it means to be healthy. Also get to grips with how different <u>recreational drugs</u>
can affect <u>behaviour</u>, <u>health</u> and <u>life processes</u>. Illegal drugs are not good news. <u>Learn it all well</u>.

Section Summary

Well, there's certainly some stuff in Section 2 — all you ever wanted to know about human beings, and a good deal more besides. Now what you've got to do is make sure you learn it all.

Remember, you have to keep coming back to these questions time and time again, to see how many of them you can do. All they do is test the basic simple facts. Let's see how much you've learnt so far...

1)   Name all five nutrients in a balanced diet.

2)   Say why each nutrient is important in the body.

3)   For each of the five nutrients, give two examples of foods they're in.

4)   Apart from the five nutrients, give two things that are needed in a balanced diet.

5)   Give two things that affect how much energy a person needs each day.

6)* Sonia has a body mass of 54 kg. What is her daily basic energy requirement?

7)   What is obesity? How is it caused?

8)   What health problems can be caused by starvation?

9)   What causes deficiency diseases?

10) What do enzymes do?

11) Name seven main bits of the digestive system.

12) Say what each of these seven bits does.

13) What are villi? What is their job? How are they well-suited to their job?

14) Give three reasons why bacteria are important in the gut.

15) What are the four jobs of the skeleton?

16) What makes bones move?

17) What are antagonistic muscles?

18) What is a bronchus? What are alveoli?

19) Name six other structures in the gas exchange system.

20) What gases are exchanged in the lungs? Where does each gas move from and to?

21) Give three ways in which the lungs are well-adapted for gas exchange.

22) What happens to the diaphragm when you breathe in?

23) What happens to the chest volume when you breathe in?
      How is this different to when you breathe out?

24) How can lung volume be measured?

25) Give two ways that regular exercise affects the gas exchange system.

26) Give three things that happen to a bronchiole during an asthma attack.

27) Give two ways in which smoking affects the gas exchange system.

28) What are the male sex cells called? Where are they made?

29) What are the female sex cells called? Where are they made?

30) What is happening when a woman has a period?

31) What starts on day 4 of the menstrual cycle? What happens at day 14?

32) Describe what an embryo looks like at:
            1 month, 9 weeks, 3 months, 5 months, 7 months, 39 weeks.

33) Explain why it's not a good idea for a woman to smoke while she's pregnant.

34) What is a 'recreational' drug?

35) Name one recreational drug. Explain how it affects behaviour.

*Answer on page 108.

Plant Nutrition

Plants make their own food — it's a nice trick if you can do it.

Photosynthesis Makes Food From Sunlight

1) Photosynthesis is a chemical process. It takes place in every green plant.
2) Photosynthesis produces food — in the form of glucose (a carbohydrate).
3) Photosynthesis happens mainly in the leaves.

Four Things are Needed for Photosynthesis...

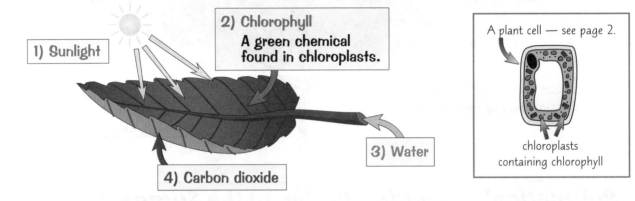

1) Sunlight

2) Chlorophyll
A green chemical found in chloroplasts.

3) Water

4) Carbon dioxide

A plant cell — see page 2.

chloroplasts containing chlorophyll

1) Chlorophyll absorbs sunlight.

2) Photosynthesis uses the energy from sunlight to turn carbon dioxide and water into glucose. Oxygen is also made.

3) You can write this word equation to show what happens:

$$\text{Carbon dioxide} + \text{Water} \xrightarrow{\text{Sunlight}} \text{Glucose} + \text{Oxygen}$$

These are the reactants.

These are the products.

There's more on reactants and products on page 45.

Leaves are Great at Photosynthesis

1) Leaves are broad. This gives them a big surface area for absorbing light.

2) Leaves have lots of chloroplasts. These are mainly at the top of the leaf, where there's most light.

3) The bottom of the leaf has lots of tiny holes called stomata.
These let carbon dioxide move into the leaf from the air. They also let oxygen move out.

Plants Also Need Things from the Soil

1) Plants need minerals from the soil to keep healthy.

2) Plants absorb minerals through their roots.

3) Plants also absorb water from the soil through their roots.

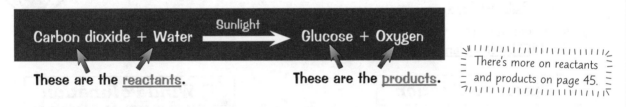

Hmm, it's all clever stuff — just make sure you learn it...

Remember, plants don't get food from the soil — they make it themselves using photosynthesis.

Plant Reproduction

Just like humans, <u>plants reproduce</u> (<u>make babies</u>). This page is all about how plant reproduction starts.

The Flower Contains the Reproductive Organs

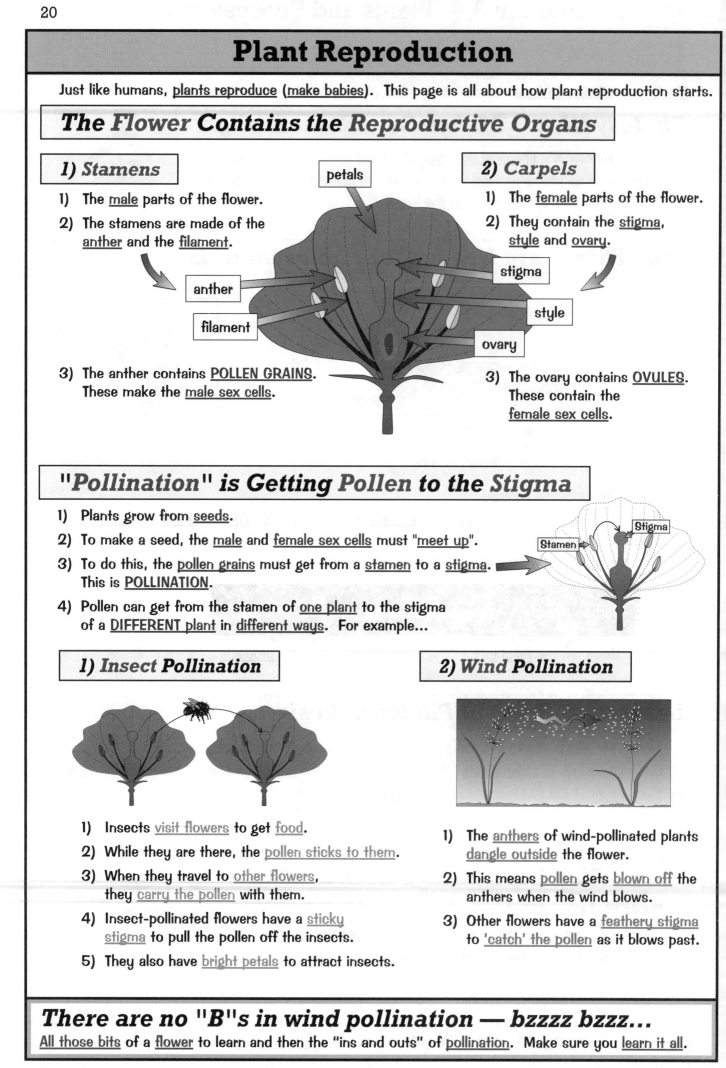

1) Stamens

1) The <u>male</u> parts of the flower.

2) The stamens are made of the <u>anther</u> and the <u>filament</u>.

petals

anther

filament

3) The anther contains <u>POLLEN GRAINS</u>. These make the <u>male sex cells</u>.

2) Carpels

1) The <u>female</u> parts of the flower.

2) They contain the <u>stigma</u>, <u>style</u> and <u>ovary</u>.

stigma

style

ovary

3) The ovary contains <u>OVULES</u>. These contain the <u>female sex cells</u>.

"Pollination" is Getting Pollen to the Stigma

1) Plants grow from <u>seeds</u>.

2) To make a seed, the <u>male</u> and <u>female sex cells</u> must "<u>meet up</u>".

3) To do this, the <u>pollen grains</u> must get from a <u>stamen</u> to a <u>stigma</u>. This is <u>POLLINATION</u>.

4) Pollen can get from the stamen of <u>one plant</u> to the stigma of a <u>DIFFERENT</u> plant in <u>different ways</u>. For example...

Stamen   Stigma

1) Insect Pollination

1) Insects <u>visit flowers</u> to get <u>food</u>.

2) While they are there, the <u>pollen sticks to them</u>.

3) When they travel to <u>other flowers</u>, they <u>carry the pollen</u> with them.

4) Insect-pollinated flowers have a <u>sticky stigma</u> to pull the pollen off the insects.

5) They also have <u>bright petals</u> to attract insects.

2) Wind Pollination

1) The <u>anthers</u> of wind-pollinated plants <u>dangle outside</u> the flower.

2) This means <u>pollen</u> gets <u>blown off</u> the anthers when the wind blows.

3) Other flowers have a <u>feathery stigma</u> to '<u>catch</u>' the pollen as it blows past.

There are no "B"s in wind pollination — bzzzz bzzz...

<u>All those bits</u> of a <u>flower</u> to learn and then the "ins and outs" of <u>pollination</u>. Make sure you <u>learn it all</u>.

Fertilisation and Seed Formation

Here's what happens __after__ a flower is __pollinated__. Make sure you've learnt the words on page 20.

Fertilisation is the Joining *of Sex Cells*

1) __Pollen__ lands on a __stigma__ with help from __insects__ or the __wind__.

2) A __pollen tube__ then grows out of a __pollen grain__ into the __ovary__.

3) The __nucleus__ from a __male sex cell__ inside the pollen grain __moves down__ the tube.

4) It __joins__ with the __nucleus__ of a __female sex cell__ inside an __ovule__. This is __FERTILISATION__.

Pollen grain

Stigma

Pollen tube

Male nucleus travelling to ovary

Female nucleus inside ovule

Seeds are Formed *From Ovules*

1) After fertilisation, the __ovule__ develops into a __seed__.

2) Each seed contains an __embryo plant__.

3) The __ovary__ develops into a __fruit__ around the seed.

A Seed:

Hard seed coat

Embryo plant

Seed Dispersal is Scattering *Seeds*

Seeds need to be __dispersed__ (scattered) from the parent plant before they can grow. Here are four __different methods__ of seed dispersal...

1) Wind dispersal

__Dandelion__ fruit.

__Parachutes__ catch the wind and carry the seeds away.

__Sycamore__ fruit.

__Wings__ help it fly away from the parent tree.

2) Animal dispersal

__Tomato__ fruit.

1) Fruit gets __eaten__.

2) Seeds come out in the animals' __poo__, away from the parent plant.

__Burdock__ fruit.

1) __Hooks__ stick to animals' coats.

2) Animals __carry__ the fruit __away__.

3) Explosions

__Peas__.

The pods __dry out__ and __flick__ the seeds out.

4) Drop and Roll

1) The heavy fruit __falls__ down from the tree.

2) It __splits__ when it hits the ground and the seeds __roll__ out.

__Horse Chestnut__ fruit.

What has a hazelnut in every bite — squirrel poo...

It all starts with __pretty flowers__. These get __pollinated__. Then __fertilisation__ happens, which makes __seeds__. Seeds then get __dispersed__. Eventually, the seeds will __grow__ into __new plants__ far away from their parents.

Investigating Seed Dispersal

At last, a little bit of <u>science in action</u>. Roll up your sleeves and let's <u>get started</u>.

You Can Investigate Seed Dispersal by Dropping Seeds

You can investigate <u>how well different seeds disperse</u>.
It's easiest to investigate <u>wind</u> and <u>drop and roll</u> dispersal.

Here's what you have to do.

1) Get a few different types of <u>fruit</u> (which contain seeds).
   For example, <u>sycamore fruit</u> and <u>horse chestnut fruit</u>.

2) Decide on a <u>fixed height</u> to drop the fruit from.

3) <u>Drop</u> the fruit <u>one at a time</u> from this height,
   directly above a <u>set point</u> on the ground.

4) <u>Measure how far</u> along the ground the seeds
   have travelled from the set point.

5) <u>Record</u> this distance in a <u>table</u>.

6) Do this experiment <u>three times</u> for each type of seed.
   Then find the <u>average distance</u> each type disperses.

Seed Type	Distance Dispersed (cm)		
Sycamore	20	25	24
Horse Chestnut			

Make Sure it's a Fair Test

You need to keep these things <u>the same</u> each time you do the experiment:

* the <u>person</u> dropping the fruit,

* the <u>height</u> the fruit are dropped from,

* the <u>place</u> you're doing the experiment (<u>stay away</u>
  from <u>doors</u> and <u>windows</u> that might cause <u>draughts</u>).

This is called "controlling the variables".

Use a Fan to Investigate the "Wind Factor"

You can investigate <u>how much</u> the <u>wind</u> affects seed dispersal using an <u>electric fan</u>.
Here's how:

1) Set up the fan a <u>fixed distance</u> from the person dropping the fruit.

2) <u>Switch the fan on</u> — it needs to be set to the <u>same speed</u> for every fruit you drop.
   This makes sure the experiment will be a <u>fair test</u>.

3) <u>Drop</u> the fruit as before and <u>measure</u> how far along the ground the seeds travel.

I've got a pea shooter — is that a seed dispersal mechanism?

You can also investigate how much the <u>shape</u> of a <u>fruit</u> helps the seeds to disperse — for example,
do the experiments above using <u>sycamore fruit</u> with the <u>wings cut off</u>. Isn't science fun?

Dependence on Other Organisms

Organisms <u>depend</u> on (need) <u>other organisms</u> for their <u>survival</u>.

Organisms *in an Ecosystem* are Interdependent

1) An <u>ecosystem</u> is <u>all</u> the <u>living organisms</u> in <u>one area</u>, plus their <u>environment</u>.

2) The <u>organisms</u> in an ecosystem are <u>interdependent</u>.

3) This means that they <u>need each other</u> to survive.

Almost All Living Things Depend on Plants

Plants Capture the Sun's Energy

1) <u>Almost all energy</u> on <u>Earth</u> comes from the <u>Sun</u>.

2) <u>Plants</u> use some of the Sun's energy to <u>make food</u> during <u>photosynthesis</u> (see page 19).

3) Plants use the food to build <u>molecules</u> (like <u>proteins</u>) which become part of the plants' cells.

4) These molecules <u>store</u> the Sun's energy.

5) The energy gets <u>passed on from plants to animals</u> when animals <u>eat</u> the plants.

6) Animals <u>can't carry out</u> photosynthesis. But they do all need <u>energy</u> to stay alive.
So animals <u>need plants</u> to capture the Sun's energy for them.

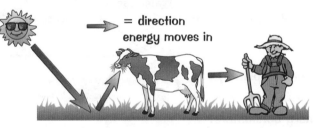

= direction
energy moves in

Plants Give Out Oxygen and Take in Carbon Dioxide

1) When plants and animals <u>respire</u> (see p. 4) they <u>TAKE IN oxygen</u> and <u>GIVE OUT carbon dioxide</u>.

2) During <u>photosynthesis</u>, plants <u>GIVE OUT oxygen</u> and <u>TAKE IN carbon dioxide</u>.

3) Without plants there <u>wouldn't be enough oxygen</u> in the air for <u>respiration</u>.

4) Also, there would be <u>too much carbon dioxide</u> in the air.

Many Plants Depend on Insects to Reproduce

1) We grow <u>crops</u> for food — for example, we grow <u>apple trees</u> to make <u>apples</u>.

2) Many crop plants need insects to <u>pollinate</u> them.

3) If they <u>don't get pollinated</u>, they <u>won't</u> be able to
make <u>fruit</u> and <u>seeds</u> for us to <u>eat</u>.

4) So we need insects to pollinate our crops and <u>give us food</u>.

We're all just one big happy family...

So. The organisms in an ecosystem are <u>interdependent</u> — we <u>depend</u> on <u>plants</u> for all our <u>energy</u> and to produce the <u>oxygen</u> we use in respiration. And many plants depend on <u>insects</u> to <u>pollinate</u> them.

Food Chains and Food Webs

Organisms mainly depend on each other for <u>food</u>.

Food Chains Show What is Eaten by What

1) This is an example of a <u>food chain</u>:

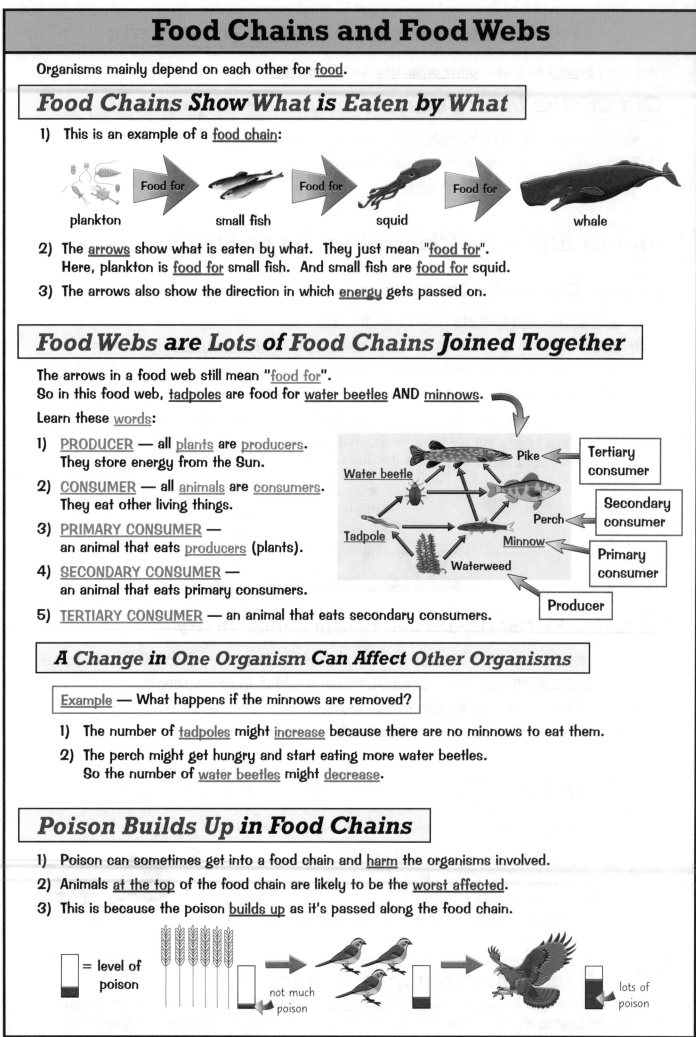

plankton   Food for   small fish   Food for   squid   Food for   whale

2) The <u>arrows</u> show what is eaten by what. They just mean "<u>food for</u>". Here, plankton is <u>food for</u> small fish. And small fish are <u>food for</u> squid.

3) The arrows also show the direction in which <u>energy</u> gets passed on.

Food Webs are Lots of Food Chains Joined Together

The arrows in a food web still mean "<u>food for</u>".
So in this food web, <u>tadpoles</u> are food for <u>water beetles</u> AND <u>minnows</u>.

Learn these <u>words</u>:

1) <u>PRODUCER</u> — all <u>plants</u> are <u>producers</u>. They store energy from the Sun.

2) <u>CONSUMER</u> — all <u>animals</u> are <u>consumers</u>. They eat other living things.

3) <u>PRIMARY CONSUMER</u> — an animal that eats <u>producers</u> (plants).

4) <u>SECONDARY CONSUMER</u> — an animal that eats primary consumers.

5) <u>TERTIARY CONSUMER</u> — an animal that eats secondary consumers.

Pike — Tertiary consumer
Water beetle
Perch — Secondary consumer
Tadpole
Minnow — Primary consumer
Waterweed — Producer

A Change in One Organism Can Affect Other Organisms

<u>Example</u> — What happens if the minnows are removed?

1) The number of <u>tadpoles</u> might <u>increase</u> because there are no minnows to eat them.

2) The perch might get hungry and start eating more water beetles. So the number of <u>water beetles</u> might <u>decrease</u>.

Poison Builds Up in Food Chains

1) Poison can sometimes get into a food chain and <u>harm</u> the organisms involved.

2) Animals <u>at the top</u> of the food chain are likely to be the <u>worst affected</u>.

3) This is because the poison <u>builds up</u> as it's passed along the food chain.

= level of poison

not much poison

lots of poison

Section Summary

The end of another section. At least plants make a change from all that gruesome human biology. No pictures of your insides here. Just lots of lovely plant diagrams to learn. Ahhh...

Make sure you know the answers to all of these questions before you turn the page.

1) What is made during photosynthesis?

2) What four things are needed for photosynthesis to happen?

3) Write the word equation for photosynthesis.

4) What two things do plants need from the soil?

5) Name the male parts of the flower. Name the female parts of the flower.

6) What is pollination?

7) Describe two ways plants can be pollinated.

8) What is a pollen tube needed for?

9) What is fertilisation?

10) What does an ovule develop into after fertilisation?

11) What does the ovary develop into?

12) What is seed dispersal?

13) Give four methods of seed dispersal.
    Give an example of a fruit that disperses seeds in each of these ways.

14) Johnny has 3 sycamore fruits. He drops each of them from a height of 1.5 metres.
    Describe what he has to do next to investigate seed dispersal.

15) How could you investigate the effect of wind on this dispersal method?

16) What is an ecosystem?

17) Where does almost all the Earth's energy come from?

18) What do plants do with this energy? How do they store it?
    Why do animals rely on plants to capture this energy?

19) Plants affect the level of oxygen in the air. Why is this important?
    What else do plants change about the air?

20) What do crops rely on insects for? How does this affect us humans?

21) What is a food chain? What is a food web?

22) What do each of the terms below mean?
    a) producer   b) consumer   c) primary consumer
    d) secondary consumer e) tertiary consumer

23)*Look at the food web on the right.
    If the number of perch decreases, the number of
    water beetles might increase. Suggest why.

24) What happens to poison as it's passed along a food chain?

*Answer on page 108.

DNA and Inheritance

This page is all about the teeny tiny things inside your cells that control what features you have.

Chromosomes, DNA and Genes

1) Most cells in your body have a nucleus.

2) The nucleus contains CHROMOSOMES. Chromosomes are long lengths of DNA.

nucleus

cell

chromosome

nucleus

DNA is a long list of chemical instructions on how to build an organism.

3) Chromosomes carry GENES. A gene is a short length of DNA.

gene

chromosome

4) Different genes control different CHARACTERISTICS (features).

eye colour gene   hair colour gene

Genes Are Passed Down From Our Parents

1) During reproduction (see p. 14) genes from the mother and father get mixed together.

2) So a baby has an equal mix of its parents' genes.

3) When genes get passed on like this it's called HEREDITY.

4) Remember genes control characteristics. So a baby will have a mixture of its parents' characteristics.

5) A characteristic passed on in this way is called a 'hereditary' characteristic.

Dad has 'big ear' gene.

'Big ear' gene passed to son by heredity.

Son has big ears — it's a hereditary characteristic.

Scientists Worked Out The Structure of DNA

1) Crick and Watson were the first scientists to build a model of DNA.

2) They used data from other scientists called Wilkins and Franklin.

3) This data helped them to understand that a DNA molecule is a spiral made of two chains twisted together.

One chain

Another chain

DNA? I get all my genes from Topshop...

There are a few tricky words to learn on this page. Test yourself by writing them all out and then writing down what they mean. Genes come up a lot in this section, so make sure you know what they are.

Variation

This page is all about <u>differences between organisms</u>. That includes <u>big, obvious differences</u>, like those between a cow and a dandelion. It also includes <u>less obvious differences</u>, like different blood groups.

Different Species Have Different Genes

1) <u>VARIATION</u> is the <u>differences</u> between living things.

2) There's variation <u>between different species</u>. This is because they have very <u>different genes</u>.

3) There's also some variation <u>within a species</u>.

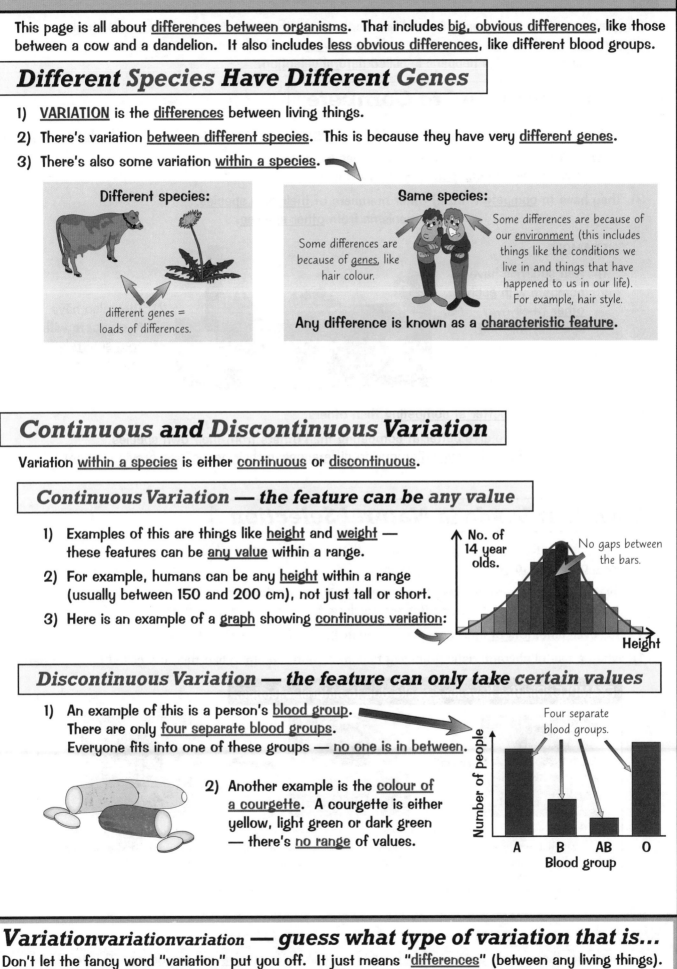

Different species:

different genes = loads of differences.

Same species:

Some differences are because of <u>genes</u>, like hair colour.

Some differences are because of our <u>environment</u> (this includes things like the conditions we live in and things that have happened to us in our life). For example, hair style.

Any difference is known as a <u>characteristic feature</u>.

Continuous and Discontinuous Variation

Variation <u>within a species</u> is either <u>continuous</u> or <u>discontinuous</u>.

Continuous Variation — the feature can be any value

1) Examples of this are things like <u>height</u> and <u>weight</u> — these features can be <u>any value</u> within a range.

2) For example, humans can be any <u>height</u> within a range (usually between 150 and 200 cm), not just tall or short.

3) Here is an example of a <u>graph</u> showing <u>continuous variation</u>:

No. of 14 year olds.

No gaps between the bars.

Height

Discontinuous Variation — the feature can only take certain values

1) An example of this is a person's <u>blood group</u>. There are only <u>four separate blood groups</u>. Everyone fits into one of these groups — <u>no one is in between</u>.

2) Another example is the <u>colour of a courgette</u>. A courgette is either yellow, light green or dark green — there's <u>no range</u> of values.

Four separate blood groups.

Number of people

A   B   AB   O

Blood group

Variationvariationvariation — guess what type of variation that is...

Don't let the fancy word "variation" put you off. It just means "<u>differences</u>" (between any living things). You can have variation <u>between</u> different species. You can also have variation <u>within</u> one species.

Natural Selection and Survival

Survive may seem pretty easy for us humans, but for other organisms it can be quite tricky. Characteristics that make organisms good at surviving are likely to become more common over time. The process by which this happens is called natural selection.

Organisms Need to Compete

1) Organisms need certain resources so they can survive and reproduce. For example, food and water.

2) Often there aren't enough of these resources to go around, so organisms need to compete ('fight') for them.

3) They have to compete with: a) other members of their own species,
   b) organisms from other species.

EXAMPLE

1) Red squirrels have to compete with each other (their own species) for food.

2) They also have to compete with grey squirrels (a different species).

4) Some species are better at competing than others.

5) Some organisms are also better at competing than others from their own species. This is because they show variation due to differences in their genes (see previous page).

Variation Leads to Natural Selection

1) Organisms with characteristics that make them better at competing are more likely to survive and reproduce.

2) This means they're more likely to pass on the genes for their useful characteristics to their offspring (children).

3) So, over time, lots of individuals end up with the useful characteristic.

4) When a useful characteristic gradually becomes more common like this, it's called natural selection.

EXAMPLE: Giraffes have long necks due to natural selection.

 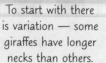 

To start with there is variation — some giraffes have longer necks than others.

Giraffes with longer necks can reach leaves easily — so they're better at competing for food. They're more likely to survive and reproduce.

The gene for a longer neck gets passed on to the next generation. This process keeps happening until all giraffes have long necks.

Well done — you've survived this page...

So, an organism that's really good at competing is more likely to survive and reproduce. The genes that made it good at competing are likely to get passed on and become more common due to natural selection.

Extinction and Preserving Species

Organisms that can't compete <u>don't survive</u> for long. It's a <u>cruel world</u> out there.

Many Species **Are at Risk of Becoming** *Extinct*

1) If the environment <u>changes</u> in some way, some organisms will be badly affected.
They may struggle to <u>compete successfully</u> for the things they need.

2) If this happens to a <u>whole species</u>, then that species may <u>die out</u>,
so there are <u>none of them left at all</u>.

3) This means the species has become <u>extinct</u> (like the woolly mammoth).

4) Species <u>at risk</u> of becoming extinct are called <u>endangered species</u>.

Humans **Can Suffer** *When Species Become* **Extinct**

1) Humans <u>rely</u> on <u>plants</u> and <u>animals</u> for loads of things. For example:

2) We need to <u>protect</u> the organisms we already use.

3) We also need to make sure organisms we <u>haven't discovered yet</u>
don't become extinct — they might end up being really important.

4) Organisms <u>rely</u> on other organisms to <u>survive</u> (see page 23).
So if one species becomes extinct, this can have a <u>knock-on effect</u>
for <u>other species</u> — including <u>humans</u>.

5) So it's important that we always have a <u>variety</u> of <u>species</u> on Earth
— this is Earth's <u>BIODIVERSITY</u>.

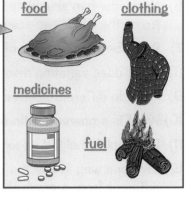

food   clothing

medicines

fuel

Gene Banks **May Help to** *Prevent Extinction*

1) A <u>gene bank</u> is a <u>store</u> of the <u>genes</u> of different species.

2) If a species becomes <u>endangered</u> or <u>extinct</u>, we could use
stored genes to <u>create new members</u> of that species.

3) So gene banks could be a way of <u>maintaining biodiversity</u> in the future.

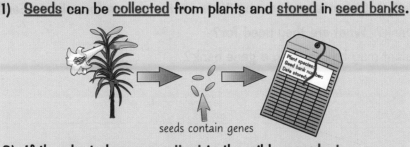

EXAMPLE: Plants

1) <u>Seeds</u> can be <u>collected</u> from plants and <u>stored</u> in <u>seed banks</u>.

seeds contain genes

2) If the plants become <u>extinct</u> in the wild, <u>new plants</u>
can be <u>grown</u> from the seeds kept in storage.

My bank balance is two seeds and an egg...

You need to understand how <u>environmental changes</u> can put species at risk of <u>extinction</u>. Also, make
sure you understand why it's important to <u>maintain biodiversity</u> and how <u>gene banks</u> may help to do this.

Section Summary

There are one or two fancy words in Section 4 which might cause you trouble.
You should make the effort to learn exactly what they mean, before you start these questions.

These questions test exactly what you know and find out exactly what you don't.
They test all the basic facts, so you need to make sure you can answer them all.
Practise these questions over and over again until you can just sail through them.

1) Where do you find chromosomes?

2) What are chromosomes made of?

3) What is a gene? What do genes control?

4) True or false? A baby gets an equal mix of its parents' genes.

5) What does heredity mean?

6) Name the two scientists who first built a model of DNA.
 Name the other two scientists whose data helped them.

7) Describe the structure of a DNA molecule.

8) What does variation mean?

9) Why do different species look different?

10) What is a characteristic feature?

11) Is 'people's height' an example of continuous or discontinuous variation? Say why.

12) Give one way in which a graph showing continuous variation would be
 different from a graph showing discontinuous variation.

13) Give two examples of resources that organisms may need to compete for.

14) Why is it important that organisms are good at competing for the things they need?

15) Why are some organisms better at competing for resources than others of the same species?

16) Why are genes for useful characteristics likely to become more common over time?
 What is this process called?

17) Why could it be bad news for an organism if its environment changes?

18) What does extinct mean?

19) What does endangered mean?

20) What is biodiversity? Give one reason why it's important for us to maintain Earth's biodiversity.

21) What is a gene bank? What are they used for?

22) What part of a plant may be stored in a gene bank?

Solids, Liquids and Gases

The first two pages in this section are all about <u>states of matter</u> and their <u>properties</u>.

The Three States of Matter — Solid, Liquid and Gas

1) Materials come in <u>three</u> different forms...

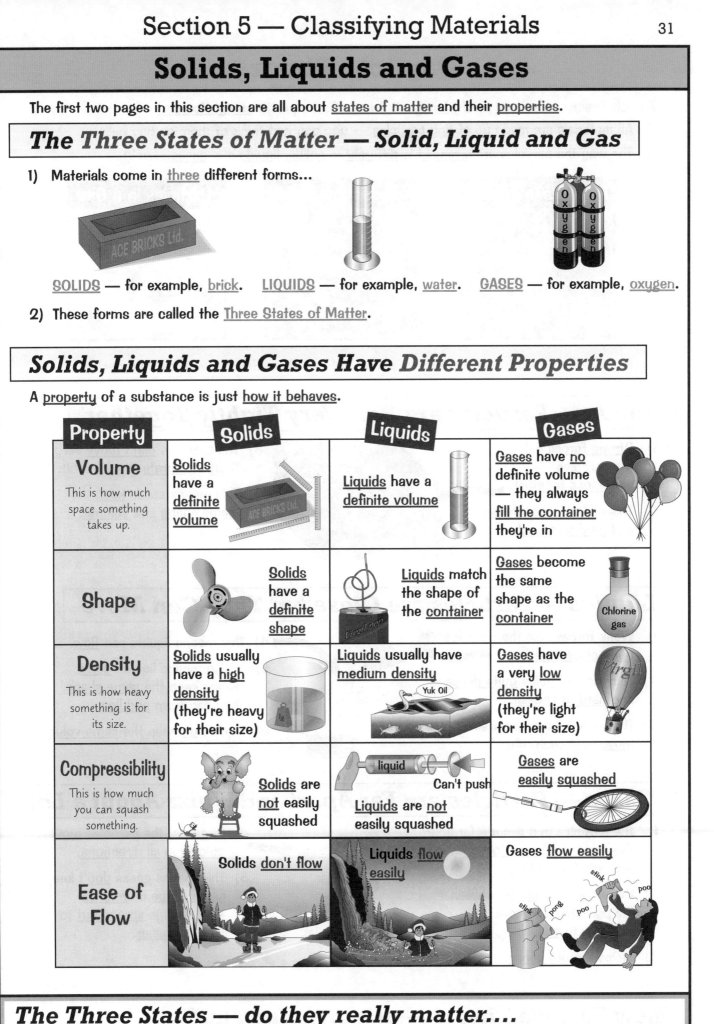

<u>SOLIDS</u> — for example, <u>brick</u>.   <u>LIQUIDS</u> — for example, <u>water</u>.   <u>GASES</u> — for example, <u>oxygen</u>.

2) These forms are called the <u>Three States of Matter</u>.

Solids, Liquids and Gases Have Different Properties

A <u>property</u> of a substance is just <u>how it behaves</u>.

Property	Solids	Liquids	Gases
Volume This is how much space something takes up.	<u>Solids</u> have a <u>definite volume</u>	<u>Liquids</u> have a <u>definite volume</u>	Gases have <u>no</u> definite volume — they always <u>fill the container</u> they're in
Shape	<u>Solids</u> have a <u>definite shape</u>	<u>Liquids</u> match the shape of the <u>container</u>	Gases become the same shape as the <u>container</u>
Density This is how heavy something is for its size.	<u>Solids</u> usually have a <u>high</u> <u>density</u> (they're heavy for their size)	<u>Liquids</u> usually have medium density	Gases have a very <u>low</u> <u>density</u> (they're light for their size)
Compressibility This is how much you can squash something.	<u>Solids</u> are <u>not</u> easily squashed	<u>Liquids</u> are <u>not</u> easily squashed	Gases are easily <u>squashed</u>
Ease of Flow	Solids <u>don't flow</u>	Liquids <u>flow easily</u>	Gases <u>flow easily</u>

The Three States — do they really matter....

<u>Solids</u>, liquids and <u>gases</u> — you must <u>learn</u> the <u>properties</u> of all three. When you think you know what they are, <u>cover the page</u> and <u>scribble</u> it all down from <u>memory</u>. Keep going until you get them all right.

Particle Theory

Particle theory — <u>sounds</u> pretty <u>fancy</u>. But actually it's pretty <u>straightforward</u>.

1) All materials are made up of <u>tiny particles</u> — you can just think of them as tiny balls.

2) The way the particles are <u>arranged</u> is <u>different</u> in <u>solids</u>, <u>liquids</u> and <u>gases</u>. Look:

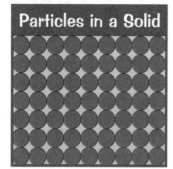
Particles in a Solid

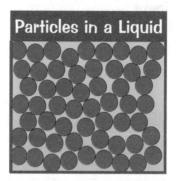
Particles in a Liquid

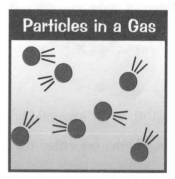
Particles in a Gas

3) <u>Particle Theory</u> explains how the arrangement of particles affects a material's <u>properties</u>.

Solids — *Particles are Held Very Tightly Together*

1) <u>Strong forces</u> hold the particles <u>very close together</u>.

2) This makes solids <u>dense</u> and <u>hard to squash</u>.

3) The particles <u>can't move</u> very much. They do <u>vibrate</u> (jiggle) a bit.

4) This means solids <u>keep</u> the <u>same shape</u> and <u>volume</u>.

Liquids — *Particles are Close but They Can Move*

1) <u>Weak forces</u> hold the particles <u>quite close together</u>.

2) This makes liquids <u>quite dense</u> and <u>hard to squash</u>.

3) The particles are also free to <u>move</u> past each other.

4) This means liquids can <u>flow</u>.

5) It also means liquids <u>don't</u> always keep the <u>same shape</u>. They can form <u>puddles</u>.

6) Liquids <u>do</u> keep the <u>same volume</u>.

Gases — *Particles are Far Apart and Whizz About a Lot*

1) The particles in a gas are <u>far apart</u>. There are <u>very weak forces</u> between the particles.

2) There's <u>lots of space</u> between the particles, so gases are <u>easy to squash</u>.

3) Gases are <u>not dense</u>.

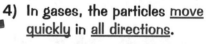

4) In gases, the particles <u>move quickly</u> in <u>all directions</u>.

5) This means gases <u>don't</u> keep the <u>same shape</u> or <u>volume</u>. They always <u>spread out</u> to <u>fill a container</u>.

Phew Particle Theory — it's gripping stuff...

It's <u>clever</u> the way you can explain all the differences between solids, liquids and gases with a page full of <u>snooker balls</u>. Anyway, that's the easy bit. The not-so-easy bit is making sure you <u>understand</u> it.

More Particle Theory

More particles I'm afraid. They do get everywhere, don't they. A bit like dog hair.

Changes of State

1) Materials can change from one state of matter to another.
   For example, water changes from a LIQUID to a SOLID when it freezes.

2) Materials change state when the arrangement and energy of the particles changes.

There's loads more on physical changes on p. 72.

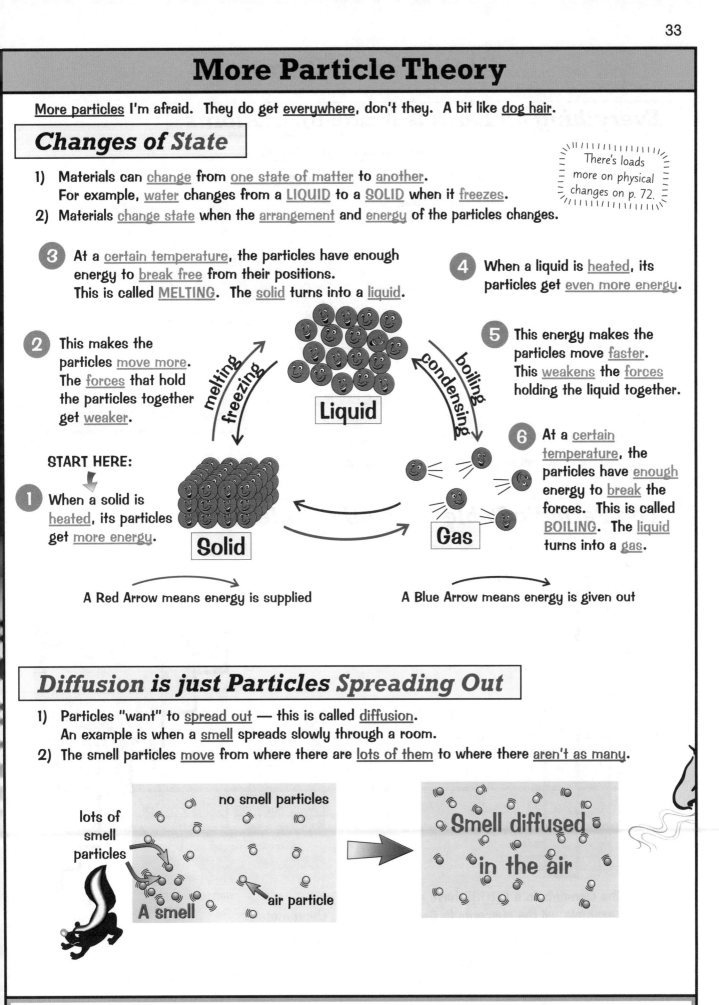

3 At a certain temperature, the particles have enough energy to break free from their positions. This is called MELTING. The solid turns into a liquid.

4 When a liquid is heated, its particles get even more energy.

2 This makes the particles move more. The forces that hold the particles together get weaker.

5 This energy makes the particles move faster. This weakens the forces holding the liquid together.

START HERE:

1 When a solid is heated, its particles get more energy.

6 At a certain temperature, the particles have enough energy to break the forces. This is called BOILING. The liquid turns into a gas.

A Red Arrow means energy is supplied

A Blue Arrow means energy is given out

Diffusion is just Particles Spreading Out

1) Particles "want" to spread out — this is called diffusion.
   An example is when a smell spreads slowly through a room.

2) The smell particles move from where there are lots of them to where there aren't as many.

lots of smell particles

no smell particles

A smell

air particle

Smell diffused in the air

Phew — another page of jostling snooker balls...

So the reason your ice cream melts is because the little snooker balls of ice cream take in energy, which means they can break free from their positions and become a liquid. Not that my ice creams last that long.

Atoms and Elements

Everything on Earth is Made up of Atoms

1) <u>ATOMS</u> are pretty much the <u>smallest</u>, <u>simplest</u> type of <u>particle</u>.

2) An <u>ELEMENT</u> is a substance that contains only <u>one type of atom</u>.

3) These substances are all elements:

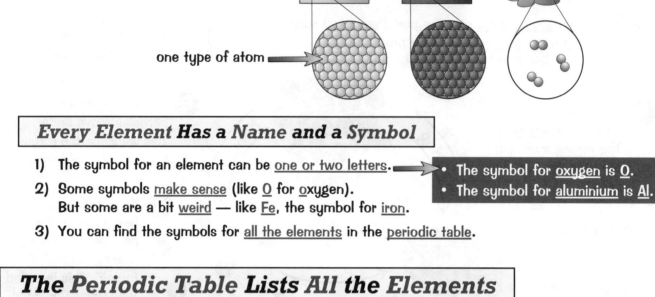

Aluminium   Iron   Oxygen

one type of atom

Every Element Has a Name and a Symbol

1) The symbol for an element can be <u>one or two letters</u>.

2) Some symbols <u>make sense</u> (like <u>O</u> for <u>oxygen</u>).
But some are a bit <u>weird</u> — like <u>Fe</u>, the symbol for <u>iron</u>.

3) You can find the symbols for <u>all the elements</u> in the <u>periodic table</u>.

- The symbol for <u>oxygen</u> is <u>O</u>.
- The symbol for <u>aluminium</u> is <u>Al</u>.

The Periodic Table Lists All the Elements

1) Elements in the periodic table are arranged in <u>GROUPS</u> and <u>PERIODS</u>.

<u>Groups</u> go <u>down</u> the table.

Group 0

	Group 1	Group 2												Group 3	Group 4	Group 5	Group 6	Group 7	
						1 H Hydrogen 1													4 He Helium 2
Period 2	7 Li Lithium 3	9 Be Beryllium 4												11 B Boron 5	12 C Carbon 6	14 N Nitrogen 7	16 O Oxygen 8	19 F Fluorine 9	20 Ne Neon 10
Period 3	23 Na Sodium 11	24 Mg Magnesium 12												27 Al Aluminium 13	28 Si Silicon 14	31 P Phosphorus 15	32 S Sulfur 16	35.5 Cl Chlorine 17	40 Ar Argon 18
Period 4	39 K Potassium 19	40 Ca Calcium 20	45 Sc Scandium 21	48 Ti Titanium 22	51 V Vanadium 23	52 Cr Chromium 24	55 Mn Manganese 25	56 Fe Iron 26	59 Co Cobalt 27	59 Ni Nickel 28	63.5 Cu Copper 29	65 Zn Zinc 30	70 Ga Gallium 31	73 Ge Germanium 32	75 As Arsenic 33	79 Se Selenium 34	80 Br Bromine 35	84 Kr Krypton 36
Period 5	86 Rb Rubidium 37	88 Sr Strontium 38	89 Y Yttrium 39					101 Ru Ruthenium 44	103 Rh Rhodium 45	106 Pd Palladium 46	108 Ag Silver 47	112 Cd Cadmium 48	115 In Indium 49	119 Sn Tin 50	122 Sb Antimony 51	128 Te Tellurium 52	127 I Iodine 53	131 Xe Xenon 54
Period 6	133 Cs Caesium 55	137 Ba Barium 56	57-71 Lanthanides	179 Hf Hafnium 72	181 Ta Tantalum 73	184 W Tungsten 74	186 Re Rhenium 75	190 Os Osmium 76	192 Ir Iridium 77	195 Pt Platinum 78	197 Au Gold 79	201 Hg Mercury 80	204 Tl Thallium 81	207 Pb Lead 82	209 Bi Bismuth 83	210 Po Polonium 84	210 At Astatine 85	222 Rn Radon 86
Period 7	223 Fr Francium 87	226 Ra Radium 88	89-103 Actinides															

<u>Periods</u> go <u>across</u>.

2) All the elements in a <u>GROUP</u> have <u>similar properties</u>.
For example, all the elements in <u>Group 1</u> are <u>soft, shiny metals</u>.

metals   non-metals   separates metals from non-metals

3) But the <u>properties</u> of elements <u>change</u> as you go <u>DOWN</u> a group.
E.g. all the Group 1 metals <u>react</u> with <u>water</u>.
But elements at the <u>bottom of the group</u> react <u>more violently</u> than elements at the <u>top</u>.

A Periodic Table for two please Sir...

You don't need to learn the name or symbol for every element. You do need to learn <u>everything else</u>.

Compounds

It would be pretty boring if we only had elements to play with.
Luckily <u>compounds</u> give us all sorts of <u>exciting materials</u>...

Compounds *Contain Two or More* Elements *Joined Up*

1) <u>Molecules</u> are formed when <u>atoms join</u>.

2) <u>Compounds</u> are formed when atoms from <u>different elements</u> join.

3) The "<u>join</u>" is known as a <u>chemical bond</u>.

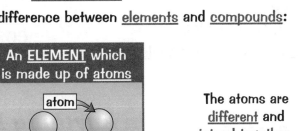

"join" or "bond" in molecule

A CO_2 molecule

You need to know the difference between <u>elements</u> and <u>compounds</u>:

An **ELEMENT** which is made up of <u>atoms</u>

atom

The atoms are <u>all</u> the <u>same</u>. It must be an <u>element</u>.

Some elements are molecules. Just remember: as long as all the atoms are the same, it's an element.

Molecules in a **COMPOUND**

The atoms are <u>different</u> and <u>joined together</u>. It must be a <u>compound</u>.

Compounds *are Formed from Chemical Reactions*

1) In a <u>chemical reaction</u> chemicals <u>combine</u> together or <u>split</u> apart to form <u>new</u> substances.

2) When a <u>new</u> compound is <u>made</u>, elements <u>combine</u>.

3) <u>New compounds</u> produced by any chemical reaction are <u>different</u> from the <u>original elements</u>. An <u>example</u> of this is <u>iron</u> reacting with <u>sulfur</u>:

Iron is <u>magnetic</u>. It reacts with <u>sulfur</u> to make <u>iron sulfide</u>.

This is a totally new substance which is <u>not magnetic</u>.

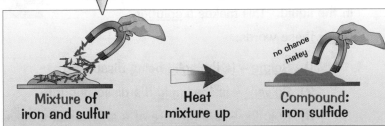

Mixture of iron and sulfur

Heat mixture up

no chance matey

Compound: iron sulfide

All Compounds *Have a Chemical Formula*

1) The formula contains the <u>symbols</u> of the <u>elements</u> that make up the compound. For example:

The symbol for <u>iron</u> is <u>Fe</u>. The symbol for <u>sulfur</u> is <u>S</u>. The <u>FORMULA</u> for <u>iron sulfide</u> is <u>FeS</u>.

2) <u>Numbers</u> in the formula tell you if there's <u>more than one atom</u> of a particular element. ➤ <u>H_2O</u> (water) has <u>two H atoms</u> and <u>one O atom</u>.

Learn about Compounds — and try and make it stick...

Teachers really do like seeing if you know the difference between elements and compounds.
It's not that tricky — but you do have to make sure you <u>learn</u> all the <u>picky details</u> on this page.

Mixtures

Mixtures in chemistry are like <u>cake mix</u> in the kitchen — all the components are <u>mushed up</u> together, but you can still <u>pick out</u> the raisins if you really want. You'll need to learn the technical terms too.

Mixtures are NOT Chemically Joined Up

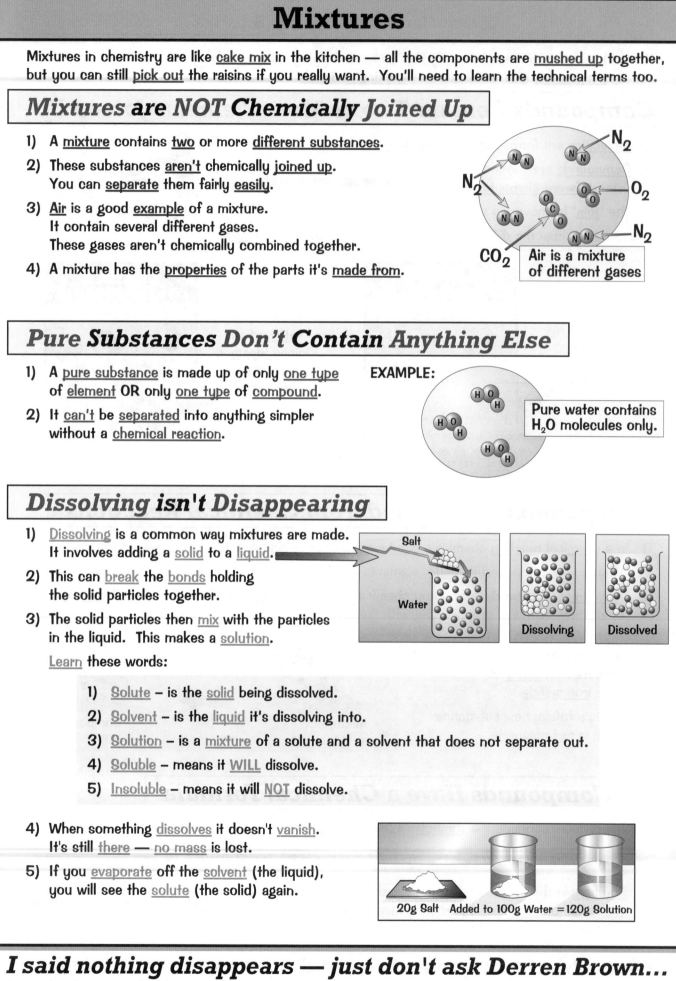

1) A <u>mixture</u> contains <u>two</u> or more <u>different substances</u>.

2) These substances <u>aren't</u> chemically <u>joined up</u>.
You can <u>separate</u> them fairly <u>easily</u>.

3) <u>Air</u> is a good <u>example</u> of a mixture.
It contain several different gases.
These gases aren't chemically combined together.

4) A mixture has the <u>properties</u> of the parts it's <u>made from</u>.

Air is a mixture
of different gases

Pure Substances Don't Contain Anything Else

1) A <u>pure substance</u> is made up of only <u>one type</u>
of <u>element</u> OR only <u>one type</u> of <u>compound</u>.

2) It <u>can't</u> be <u>separated</u> into anything simpler
without a <u>chemical reaction</u>.

EXAMPLE:

Pure water contains
H_2O molecules only.

Dissolving isn't Disappearing

1) <u>Dissolving</u> is a common way mixtures are made.
It involves adding a <u>solid</u> to a <u>liquid</u>.

2) This can <u>break</u> the <u>bonds</u> holding
the solid particles together.

3) The solid particles then <u>mix</u> with the particles
in the liquid. This makes a <u>solution</u>.

<u>Learn</u> these words:

Salt

Water

Dissolving

Dissolved

    1) <u>Solute</u> – is the <u>solid</u> being dissolved.

    2) <u>Solvent</u> – is the <u>liquid</u> it's dissolving into.

    3) <u>Solution</u> – is a <u>mixture</u> of a solute and a solvent that does not separate out.

    4) <u>Soluble</u> – means it <u>WILL</u> dissolve.

    5) <u>Insoluble</u> – means it will <u>NOT</u> dissolve.

4) When something <u>dissolves</u> it doesn't <u>vanish</u>.
It's still <u>there</u> — <u>no mass</u> is lost.

5) If you <u>evaporate</u> off the <u>solvent</u> (the liquid),
you will see the <u>solute</u> (the solid) again.

20g Salt   Added to 100g Water =120g Solution

I said nothing disappears — just don't ask Derren Brown...

<u>Learn</u> the three main headings on this page till you can write them down <u>from memory</u>.
Then <u>learn</u> the stuff underneath them, including the diagrams. <u>Cover the page</u> and <u>write it all down</u>.

Separating Mixtures

There are all sorts of ways you can separate mixtures. You've got to know <u>four</u> of them.

Mixtures *Can be Separated Using Physical Methods*

There are <u>four separation methods</u> you need to know:

1) <u>FILTRATION</u>   2) <u>EVAPORATION</u>   3) <u>CHROMATOGRAPHY</u>   4) <u>DISTILLATION</u> (see next page).

1) Filtration and 2) Evaporation — E.g. for the Separation of Rock Salt

1) <u>Rock Salt</u> is a <u>mixture</u> of <u>salt</u> and <u>sand</u>.

2) Salt and sand are both <u>compounds</u> — but <u>salt dissolves</u> in water and <u>sand doesn't</u>.

3) This <u>difference</u> allows us to <u>separate</u> them.

Learn the Four Steps of the Method:

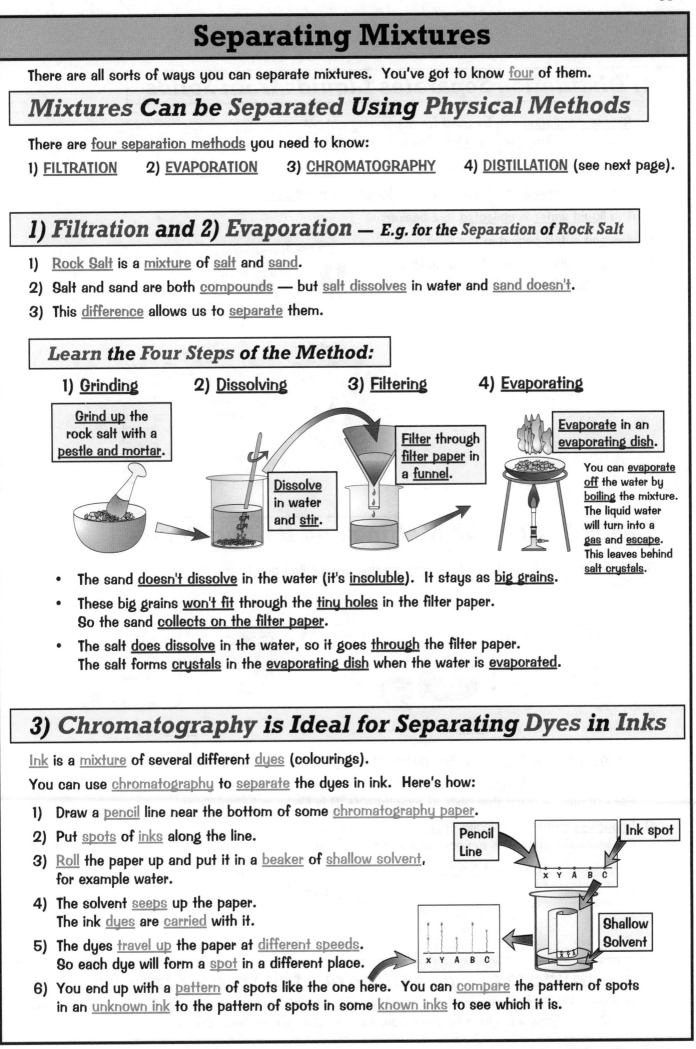

1) <u>Grinding</u>

<u>Grind up</u> the rock salt with a <u>pestle and mortar.</u>

2) <u>Dissolving</u>

<u>Dissolve</u> in water and <u>stir</u>.

3) <u>Filtering</u>

<u>Filter</u> through <u>filter paper</u> in a <u>funnel</u>.

4) <u>Evaporating</u>

<u>Evaporate</u> in an <u>evaporating dish</u>.

You can <u>evaporate</u> <u>off</u> the water by <u>boiling</u> the mixture. The liquid water will turn into a <u>gas</u> and <u>escape</u>. This leaves behind <u>salt crystals</u>.

• The sand <u>doesn't dissolve</u> in the water (it's <u>insoluble</u>). It stays as <u>big grains</u>.

• These big grains <u>won't fit</u> through the <u>tiny holes</u> in the filter paper. So the sand <u>collects on the filter paper</u>.

• The salt <u>does dissolve</u> in the water, so it goes <u>through</u> the filter paper. The salt forms <u>crystals</u> in the <u>evaporating dish</u> when the water is <u>evaporated</u>.

3) Chromatography is Ideal for Separating Dyes in Inks

Ink is a <u>mixture</u> of several different <u>dyes</u> (colourings).

You can use <u>chromatography</u> to <u>separate</u> the dyes in ink. Here's how:

1) Draw a <u>pencil</u> line near the bottom of some <u>chromatography paper</u>.

2) Put <u>spots</u> of <u>inks</u> along the line.

3) <u>Roll</u> the paper up and put it in a <u>beaker</u> of <u>shallow solvent</u>, for example water.

4) The solvent <u>seeps</u> up the paper. The ink <u>dyes</u> are <u>carried</u> with it.

5) The dyes <u>travel up</u> the paper at <u>different speeds</u>. So each dye will form a <u>spot</u> in a different place.

6) You end up with a <u>pattern</u> of spots like the one here. You can <u>compare</u> the pattern of spots in an <u>unknown ink</u> to the pattern of spots in some <u>known inks</u> to see which it is.

Pencil Line

Ink spot

X Y A B C

Shallow Solvent

X Y A B C

Separating Mixtures

4) Distillation Separates Liquids from Solids

1) Simple distillation can be used for separating out a mixture of a liquid and a solid. For example, salt water.

2) The mixture is heated in a flask and the water boils off to form a gas.

3) The gas is cooled and turns back to a liquid in a condenser.

4) The liquid water is collected in a beaker.

5) The salt is left behind in the flask.

6) Simple distillation can also be used to get pure water from dirty water.

Thermometer
Cooling water out
Flask
Salt water
Condenser
Cooling water in
Pure distilled water
Beaker
Heat

Check Purity with Melting and Boiling Points

1) A pure chemical substance has fixed melting and boiling points. For example:

Pure water boils at 100 °C.

Pure ice melts at 0 °C.

2) We know the melting and boiling points of a huge range of substances.

3) This helps us to identify substances if we're not sure what they are. For example, a liquid that boils at exactly 100 °C is likely to be pure water.

4) Impurities (other chemicals) change melting and boiling points. For example, impurities in water cause it to boil above 100 °C.

5) This means you can test the purity of a substance you've separated from a mixture. So if you want to test the purity of some water, boil it.

Revise mixtures — just filter out the important bits...

Teachers love asking you about separation techniques, so make sure you can scribble down all the stuff on pages 37-38. Draw a diagram to show how each method works too. Enjoy.

Properties of Metals

Metals are jolly <u>useful</u>. We use them all the time in <u>wires</u>, <u>bridges</u>, <u>musical instruments</u> and more. So it's only fair that you learn these two pages of fab <u>facts</u> about them in return...

1) Metals *Can be Found in the Periodic Table*

1) <u>Most</u> of the elements in the periodic table are metals.

2) Some are shown here in red, to the <u>left</u> of the <u>zig zag</u>.

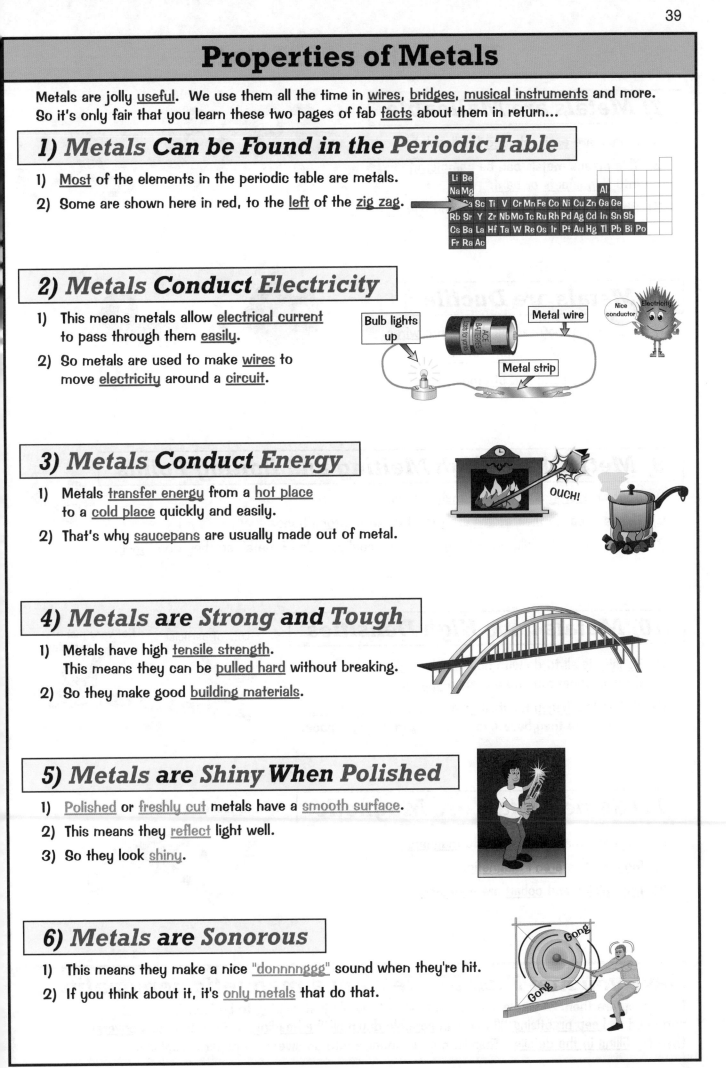

2) Metals *Conduct Electricity*

1) This means metals allow <u>electrical current</u> to pass through them <u>easily</u>.

2) So metals are used to make <u>wires</u> to move <u>electricity</u> around a <u>circuit</u>.

3) Metals *Conduct Energy*

1) Metals <u>transfer energy</u> from a <u>hot place</u> to a <u>cold place</u> quickly and easily.

2) That's why <u>saucepans</u> are usually made out of metal.

4) Metals *are Strong and Tough*

1) Metals have high <u>tensile strength</u>. This means they can be <u>pulled hard</u> without breaking.

2) So they make good <u>building materials</u>.

5) Metals *are Shiny When Polished*

1) <u>Polished</u> or <u>freshly cut</u> metals have a <u>smooth surface</u>.

2) This means they <u>reflect</u> light well.

3) So they look <u>shiny</u>.

6) Metals *are Sonorous*

1) This means they make a nice "<u>donnnnggg</u>" sound when they're hit.

2) If you think about it, it's <u>only metals</u> that do that.

Properties of Metals

7) Metals are Malleable

1) Metals are easily shaped (malleable).

2) This means metals can be hammered into thin sheets or bent.

3) Shaped metal is used in cars and planes.

8) Metals are Ductile

1) This means they can be drawn into wires.

2) This means metals aren't brittle like non-metals are (see page 42).

3) Metals just bend and stretch.

9) Metals have High Melting and Boiling Points

1) A lot of energy is needed to melt metals.

2) This is because their atoms are joined up with strong bonds.

3) Things that get really hot (like ovens) are usually made of metal, so they don't melt.

10) Metals have High Densities

1) Density is all to do with how much stuff there is squeezed into a certain space.

2) Metals feel heavy for their size — they're very dense. It's because they have a lot of atoms in a small space.

Lots of particles

Not very many particles

Metal

Non-metal

11) Some Metals are Magnetic

Iron or nickel or cobalt

1) Magnetic means attracted to magnets.

2) Most metals aren't magnetic.

3) Iron, nickel and cobalt are magnetic.

Have you met Nick L? He's got a magnetic personality...

There they are then. A whole load of facts about metals just waiting to be learnt.

You need to keep practising till you can scribble down all the headings with both pages covered.

Then try filling in the details. Then turn your doodles into an awesome paper aeroplane.

Properties of Non-Metals

The properties of non-metal elements <u>vary</u> a lot. Good — life would stink if everything was like sulfur...

1) Non-metals *Can be Found in the Periodic Table*

1) Non-metals are on the <u>right</u> of the <u>zig zag</u>.
   Look, right over there.

2) There are <u>fewer</u> non-metals than metals.

2) Non-metals *Don't Conduct Electricity*

1) Most non-metals are <u>insulators</u>.

2) This means electrical current <u>can't</u> flow through them.

3) This is <u>useful</u> — non-metals are used to make things like <u>plugs</u> and electric cable <u>coverings</u>.

One exception to this rule is <u>graphite</u>.
It's a <u>non-metal</u> made of <u>carbon</u>.
But graphite <u>can conduct electricity</u>.

3) Non-metals *Don't Conduct Energy by Heating Well*

1) Non-metals <u>don't</u> transfer energy from a <u>hot place</u> to a <u>cold place</u> quickly or easily.

2) This makes non-metals really good <u>insulators</u>.

3) <u>Oven gloves</u>, <u>saucepan handles</u> and <u>loft insulation</u> are normally made of <u>non-metals</u>.

4) Non-metals *are NOT Strong or Hard-Wearing*

1) The <u>forces</u> between the particles in non-metals are <u>weak</u>.

2) This means non-metals <u>break</u> easily.

3) It's also easy to <u>scrub</u> atoms or molecules off them — so they <u>wear away</u> quickly.

This carbon brush keeps wearing out

Properties of Non-Metals

5) Non-metals are Dull

1) Most non-metals don't <u>reflect</u> light well.
2) Their surfaces are not usually as <u>smooth</u> as metals.
3) This makes them look <u>dull</u>.

6) Non-metals are Brittle

1) Non-metal <u>structures</u> are held together by <u>weak forces</u>.
2) This means they can <u>shatter</u> all too easily.

7) Non-metals Have Low Melting Points and Boiling Points

1) The <u>forces</u> which hold the particles in non-metals <u>together</u> are <u>very weak</u>.
   This means they <u>melt</u> and <u>boil</u> very <u>easily</u>.
2) At <u>room temperature</u>, most non-metals are <u>gases</u> or <u>solids</u>. Only one is <u>liquid</u>.

8) Non-metals Have Low Densities

1) This means they <u>don't</u> have many <u>particles</u> packed into a certain <u>space</u>.
2) Non-metals which are <u>gases</u> will have <u>very low density</u>.
   Some of these gases will even <u>float</u> in <u>air</u>.
   These are ideal for <u>party balloons</u>.
3) Even the liquid and solid non-metals have <u>low densities</u>.

9) Non-metals are Not Magnetic

1) Only a few <u>metals</u> (iron, cobalt and nickel) are <u>magnetic</u>.
2) <u>All non-metals</u> are definitely <u>non-magnetic</u>.

Non-Metals — they REALLY ARE dull aren't they...

You still have to learn all about them though. Do it like this: <u>cover the page</u> with a bit of paper and try and <u>write down</u> each of the 9 points, one at a time. Lower the paper each time to see if you scribbled it all down right. Keep trying <u>till you can get them all</u>. Then put on some relaxing music by Non-Metallica.

Properties of Other Materials

As well as metals and non-metals, you need to learn all about some <u>compounds</u> and <u>mixtures of compounds</u> — <u>polymers</u>, <u>ceramics</u> and <u>composites</u>.

Polymers *Have Many Useful Properties*

Another name for polymers is <u>plastics</u>. Nylon, polythene and PVC are all polymers.

1) Polymers are usually <u>insulators</u>. It's difficult for <u>energy</u> to be transferred through them <u>electricallly</u> or <u>by heating</u>.

2) They're often <u>flexible</u>. They can be <u>bent</u> without breaking.

3) They can be very <u>light</u> for their size and strength. So they're ideal for making things that need to be <u>strong</u> but <u>not heavy</u>.

Polymers are used to make everything from <u>kayaks</u> to <u>carrier bags</u>.

4) They're <u>easily moulded</u>. They can be used to manufacture equipment with almost <u>any shape</u>.

Ceramics *are Stiff but Brittle*

Ceramics include <u>glass</u>, <u>porcelain</u> and <u>bone china</u> (for posh tea cups). They are:

1) <u>Insulators</u> of <u>heat</u> and <u>electricity</u>.

2) <u>Brittle</u> — they aren't very <u>flexible</u> and will <u>break</u> instead of <u>bending</u>.

As well as <u>tea cups</u>, ceramics are used for <u>brakes</u> in cars.

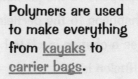

3) <u>Stiff</u> — they can withstand strong forces before they break.

Composites *are Made of Different Materials*

1) <u>Composite materials</u> are made from <u>two or more materials</u> stuck together.

2) This can make a material with <u>more useful</u> properties than either material alone. For example:

Fibreglass

1) <u>Fibreglass</u> is made up of <u>glass fibres</u> fixed in <u>plastic</u>.
2) It has a <u>low density</u> (like plastic) but is <u>very strong</u> (like glass).
3) So fibreglass is used to make <u>skis</u>, <u>boats</u> and <u>surfboards</u>.

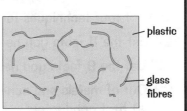

plastic

glass fibres

fibreglass

Concrete

1) <u>Concrete</u> is made from a mixture of <u>sand</u> and <u>gravel</u> mixed in <u>cement</u>.
2) It can cope with being <u>squashed</u> without breaking. So it's great at supporting heavy things.
3) This makes it ideal for use as a <u>building material</u>, for example in skate parks, shopping centres, and airports.

concrete

Boredom is a common property of revision...

Polymers, ceramics and composites. They're pretty handy — especially when the alternative is a wooden surfboard. Now get learning <u>everything</u> on this page. Don't stop until you can describe all 3 materials <u>AND</u> their uses without any <u>sneaky peeking</u>. Then stare smugly out the window for 30 seconds.

Section Summary

We've moved on to Chemistry now. Makes a refreshing change from all that slimy Biology anyway.

You know the drill: work through these questions and try to answer them.
For any you can't do, look back through Section 5 and find the answer — and then learn it.
Then try all the questions again and see how many more you can do that time.

1) What are the three states of matter? Describe two properties for each of them.

2) Draw what the particles look like in a solid, a liquid and a gas.

3) Describe what happens to the particles in a solid when the solid melts.

4) Describe what happens to the particles in a liquid when it changes into a gas.

5) What effect does a change in state have on mass?

6) Explain what diffusion is.

7) What is an atom?

8) What is an element?

9) Do groups go down the periodic table or across?

10) Using the periodic table, give the chemical symbol for these:
    a) sodium   b) magnesium   c) oxygen   d) iron   e) sulfur
    f) aluminium   g) carbon   h) chlorine   i) calcium   j) zinc.

11) True or false? All elements in a group have similar properties.

12) What is a compound?

13) Sketch some molecules that could be in a compound.

14) In what way is iron sulfide different from a mixture of iron and sulfur?

15) Which two elements are in the compound H_2O? (Use the periodic table on page 34 to help you.)

16)*Which two elements are in the compound $CaCl$? (Use the periodic table on page 34 to help you.)

17) What is a mixture?

18) What is a pure substance?

19) Describe what happens when a substance dissolves.

20) List four techniques for separating a mixture. Give an example of when you would use each one.

21) List the 11 facts you need to know about metals.

22) List the 9 facts you need to know about non-metals.

23) Out of metals and non-metals, which are the:
    a) best conductors   b) most brittle   c) strongest   d) best insulators?

24) Name 4 useful properties of polymers.

25) What are ceramics useful for?

26) What are composites? Name one and describe what it's made of.

*Answer on page 108

Equations

Chemistry is full of equations. Luckily, you'll find all you need to know about them here.

Chemicals are Formed in Chemical Reactions

1) In a chemical reaction, chemicals combine together or split apart to form new substances.

2) The chemicals you start with are called the REACTANTS.

3) The chemicals you end up with are called the PRODUCTS.

Word Equations Show What's Happening in a Reaction

1) You can show what's happening in a chemical reaction by writing an equation.

2) A word equation has the names of all the chemicals written out in full.

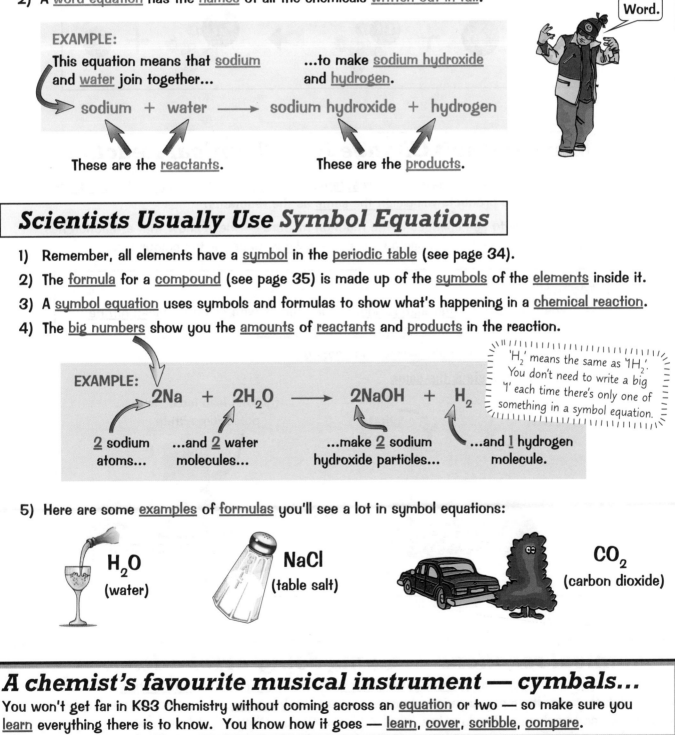

EXAMPLE:

This equation means that sodium and water join together...

...to make sodium hydroxide and hydrogen.

sodium + water ⟶ sodium hydroxide + hydrogen

These are the reactants.

These are the products.

Word.

Scientists Usually Use Symbol Equations

1) Remember, all elements have a symbol in the periodic table (see page 34).

2) The formula for a compound (see page 35) is made up of the symbols of the elements inside it.

3) A symbol equation uses symbols and formulas to show what's happening in a chemical reaction.

4) The big numbers show you the amounts of reactants and products in the reaction.

EXAMPLE: $2Na + 2H_2O \longrightarrow 2NaOH + H_2$

2 sodium atoms... ...and 2 water molecules... ...make 2 sodium hydroxide particles... ...and 1 hydrogen molecule.

'H_2' means the same as '$1H_2$'. You don't need to write a big '1' each time there's only one of something in a symbol equation.

5) Here are some examples of formulas you'll see a lot in symbol equations:

H_2O (water) $NaCl$ (table salt) CO_2 (carbon dioxide)

A chemist's favourite musical instrument — cymbals...

You won't get far in KS3 Chemistry without coming across an equation or two — so make sure you learn everything there is to know. You know how it goes — learn, cover, scribble, compare.

Chemical Reactions

In a chemical reaction, all that's really happening is the <u>atoms are moving around</u> into new positions. The reactants might give out energy or make a loud bang, but the <u>mass won't change</u>.

Atoms Move About During Chemical Reactions

1) In a <u>chemical reaction</u> atoms are <u>not</u> made or destroyed.

2) The atoms <u>move around</u> during a chemical reaction, but they're <u>not changed</u>.

3) The atoms at the <u>start</u> of a reaction are <u>still there</u> at the <u>end</u>.

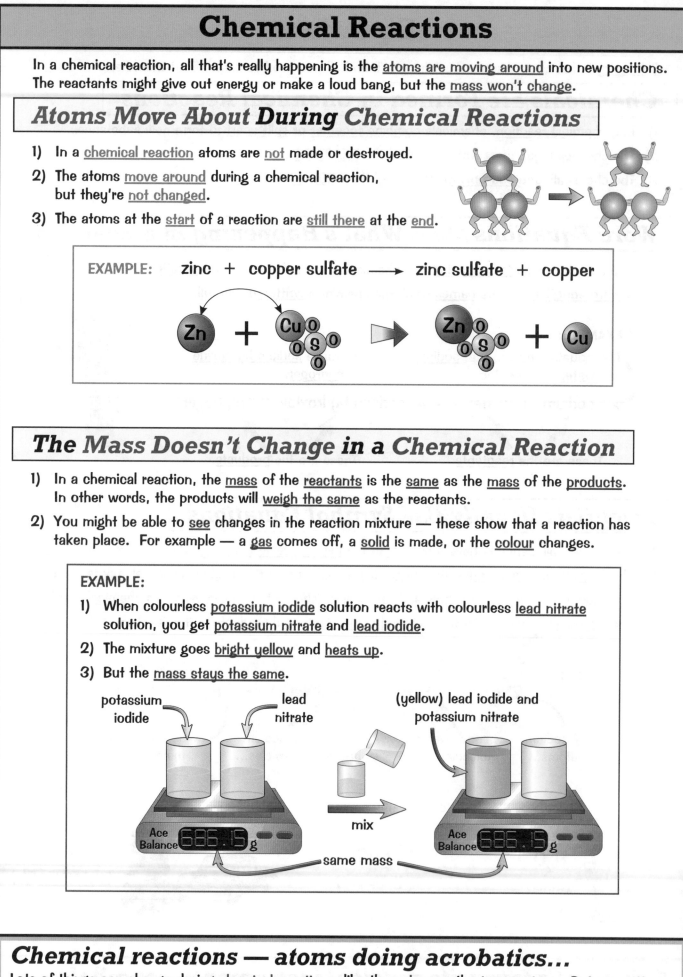

EXAMPLE: zinc + copper sulfate ⟶ zinc sulfate + copper

The Mass Doesn't Change in a Chemical Reaction

1) In a chemical reaction, the <u>mass</u> of the <u>reactants</u> is the <u>same</u> as the <u>mass</u> of the <u>products</u>. In other words, the products will <u>weigh the same</u> as the reactants.

2) You might be able to <u>see</u> changes in the reaction mixture — these show that a reaction has taken place. For example — a <u>gas</u> comes off, a <u>solid</u> is made, or the <u>colour</u> changes.

EXAMPLE:

1) When colourless <u>potassium iodide</u> solution reacts with colourless <u>lead nitrate</u> solution, you get <u>potassium nitrate</u> and <u>lead iodide</u>.

2) The mixture goes <u>bright yellow</u> and <u>heats up</u>.

3) But the <u>mass stays the same</u>.

potassium iodide

lead nitrate

(yellow) lead iodide and potassium nitrate

mix

Ace Balance 686.15 g

Ace Balance 686.15 g

same mass

Chemical reactions — atoms doing acrobatics...

Lots of things can change during <u>chemical reactions</u>, like the <u>colour</u> or the <u>temperature</u>. But no matter what else changes, the <u>mass</u> of the products will always be <u>the same</u> as the mass of the reactants. Learn that and you've practically solved all of KS3 chemistry. Okay, maybe not quite.

Examples of Chemical Reactions

Three examples of <u>chemical reactions</u> coming right up... just what the doctor ordered.

Combustion *is Burning in Oxygen*

1) Combustion is <u>burning</u> — a <u>fuel</u> reacts with <u>oxygen</u> to release <u>energy</u>.

2) <u>Three</u> things are needed for combustion:

> 1) Fuel
> 2) Heating
> 3) Oxygen

3) <u>Hydrocarbons</u> are <u>fuels</u> containing only <u>hydrogen</u> and <u>carbon</u>.

4) When it's <u>hot</u> enough and there's enough <u>oxygen</u>, hydrocarbons <u>combust</u> (burn) to give <u>water</u> and <u>carbon dioxide</u>:

> hydrocarbon + oxygen ⟶ carbon dioxide + water (+ energy)

5) Combustion is useful because <u>energy</u> is transferred away by <u>light</u> and by <u>heating</u>.

Oxidation *is the Gain of Oxygen*

1) When a substance <u>reacts</u> with <u>oxygen</u>, it's called an <u>oxidation</u> reaction.

2) <u>Combustion</u> is an oxidation reaction.

3) <u>Rusting</u> is also an oxidation reaction.

4) Rusting is when <u>iron</u> reacts with <u>oxygen</u> in the air to form <u>iron oxide</u> (<u>rust</u>).

> iron + oxygen ⟶ iron oxide (rust)

Thermal Decomposition *is Breaking Down by Heating*

1) <u>Thermal decomposition</u> is when you <u>heat</u> a substance and it <u>breaks down</u>.

2) For example, if you heat <u>copper carbonate</u> it <u>breaks down</u> into copper oxide and carbon dioxide.

> copper carbonate ⟶ copper oxide + carbon dioxide
> $CuCO_3$ CuO + CO_2

This is <u>green</u>...   ...and this is <u>black</u>.

3) Only <u>certain</u> substances <u>break down</u> when they're heated — usually things just <u>melt</u>.

It's too hot... I'm breaking down...

This page is easy — let me break it down for you...

Here are three types of <u>chemical reaction</u> to read up and <u>learn</u>. They're here because they're important, so make sure you learn them properly. That way you'll have plenty of time to enjoy what's coming up on the next page — that's right, yet more fun and interesting chemical reactions.

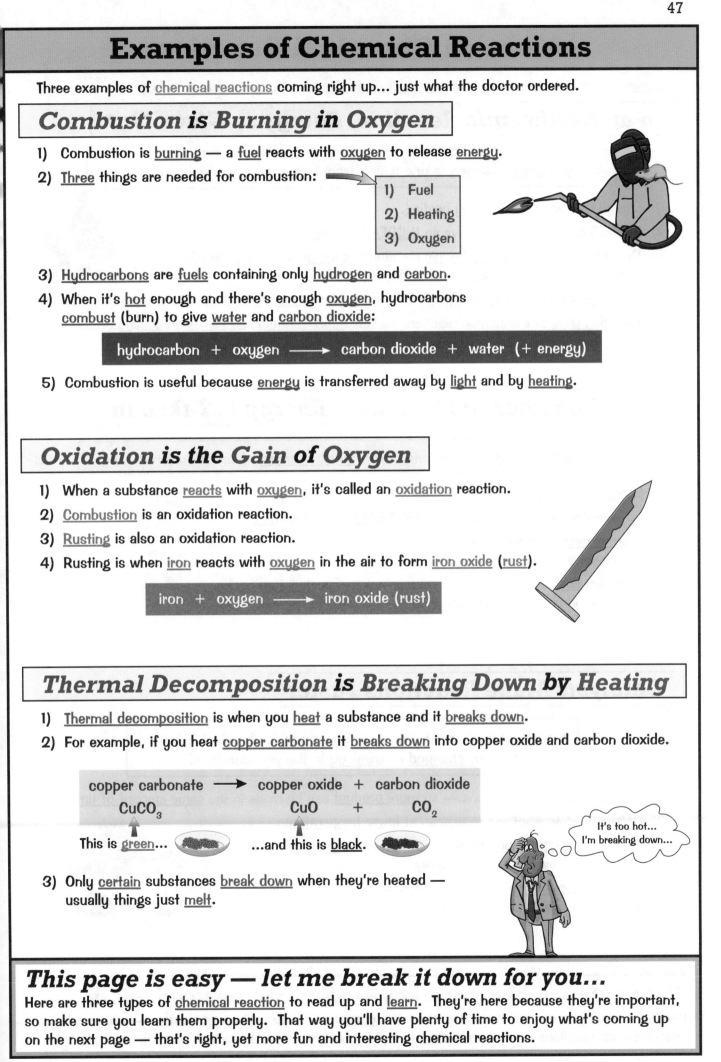

More on Chemical Reactions

Chemical reactions always involve a transfer of energy to or from the surroundings.

In an Exothermic Reaction, Energy is Given Out

> An exothermic reaction is one which transfers energy to the surroundings.

1) Energy is usually given out by heating.
2) So exothermic reactions involve an INCREASE in temperature.
3) The best example of an exothermic reaction is combustion (see page 47). This gives out a lot of energy — it's very exothermic.
4) Many neutralisation reactions (page 50) and oxidation reactions (page 47) are exothermic.
5) Everyday uses of exothermic reactions include hand warmers and self-heating cans of coffee.

In an Endothermic Reaction, Energy is Taken in

> An endothermic reaction is one which takes in energy from the surroundings.

1) Energy is usually taken in by heating.
2) So endothermic reactions involve a DECREASE in temperature.
3) Thermal decompositions (page 47) are endothermic reactions. They involve a substance taking in energy, then breaking down.
4) Everyday uses of endothermic reactions include sports injury packs. They take in energy and get very cold.

Catalysts Make Reactions Faster

> A catalyst is a substance which speeds up a chemical reaction. It is not changed or used up in the reaction itself.

1) Catalysts speed up reactions — so more product can be made in the same amount of time.
2) Catalysts allow reactions to happen at lower temperatures.
3) In industry, high temperatures make reactions expensive to run. So catalysts make reactions cheaper.
4) Catalysts come out of a reaction the same as when they went in. This means catalysts can be reused.

Catalysts are like my jokes — they can be used again...

Catalysts aren't just used in chemistry labs. The enzymes in your body are catalysts — without them, your chemical reactions would be too slow to keep you alive. There's more on enzymes on p. 8.

Acids and Alkalis

The pH scale is what scientists use to describe how acidic or alkaline a substance is.

The pH Scale Shows the Strength of Acids and Alkalis

1) The pH scale goes from 0 to 14.

2) Anything with a pH below 7 is an acid. The strongest acid has pH 0.

3) Anything with a pH above 7 is an alkali. The strongest alkali has pH 14.

4) A neutral substance has pH 7 (like water).

Indicators Are Dyes Which Change Colour

1) An indicator is something that changes colour depending on whether it's in an acid or in an alkali.

2) Litmus paper is an indicator. Acids turn litmus paper red. Alkalis turn it blue.

3) Universal indicator solution is a liquid indicator.

4) It gives the colours shown in a pH chart.

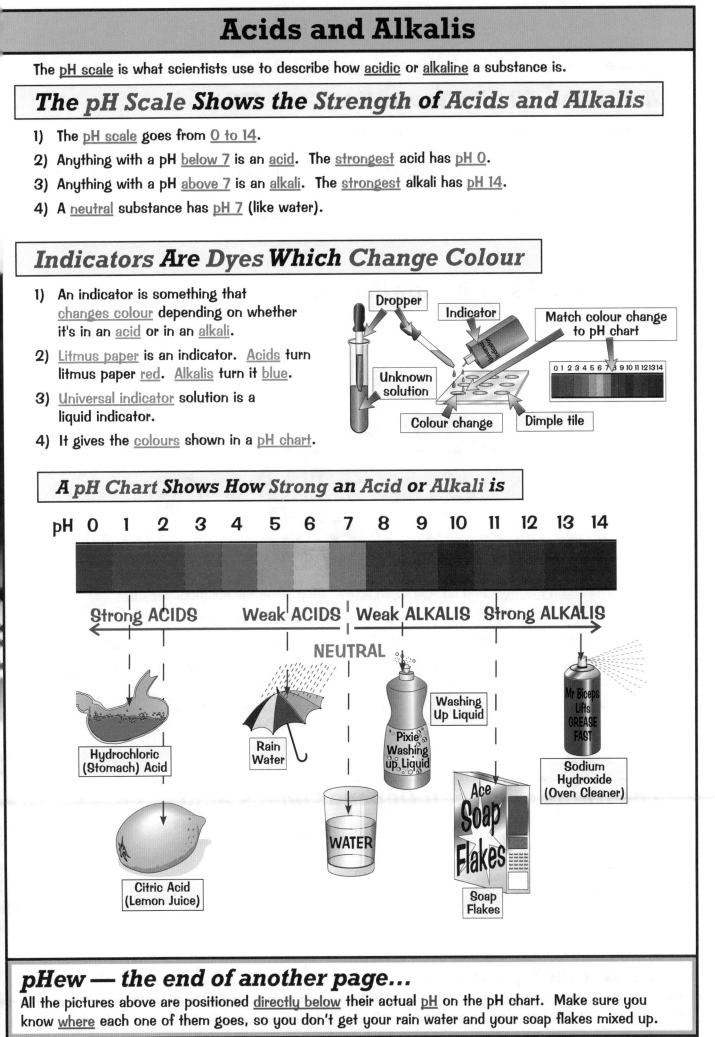

Dropper

Indicator

Match colour change to pH chart

0 1 2 3 4 5 6 7 8 9 10 11 12 13 14

Unknown solution

Colour change

Dimple tile

A pH Chart Shows How Strong an Acid or Alkali is

pH 0   1   2   3   4   5   6   7   8   9   10   11   12   13   14

Strong ACIDS   Weak ACIDS   Weak ALKALIS   Strong ALKALIS

NEUTRAL

Hydrochloric (Stomach) Acid

Rain Water

Washing Up Liquid

Pixie Washing up Liquid

Mr Biceps Lifts GREASE FAST

Sodium Hydroxide (Oven Cleaner)

Citric Acid (Lemon Juice)

WATER

Ace Soap Flakes

Soap Flakes

pHew — the end of another page...

All the pictures above are positioned directly below their actual pH on the pH chart. Make sure you know where each one of them goes, so you don't get your rain water and your soap flakes mixed up.

Neutralisation Reactions

You might have done something like this in the <u>lab</u>. If not, I bet you will pretty soon.

Acids and Alkalis Neutralise Each Other

1) <u>Acids</u> react with <u>alkalis</u> to form a <u>salt</u> and <u>water</u>:

$$\text{acid} + \text{alkali} \longrightarrow \text{salt} + \text{water}$$

> You can get different kinds of salt — not just table salt.

2) This is a <u>neutralisation</u> reaction. The products have a <u>neutral pH</u> (a pH of 7).

Making Salts by Neutralisation

Making <u>salts</u> is pretty easy — you just need a steady hand and a lot of time. A bit like whisking eggs.

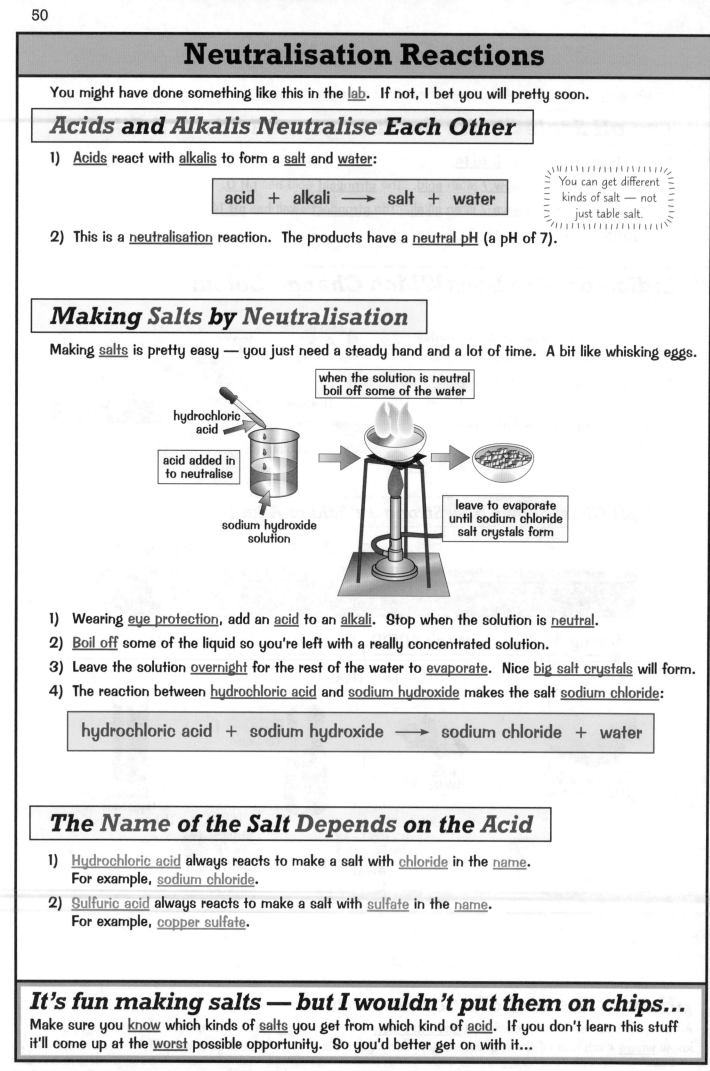

hydrochloric acid

acid added in to neutralise

sodium hydroxide solution

when the solution is neutral boil off some of the water

leave to evaporate until sodium chloride salt crystals form

1) Wearing <u>eye protection</u>, add an <u>acid</u> to an <u>alkali</u>. Stop when the solution is <u>neutral</u>.

2) <u>Boil off</u> some of the liquid so you're left with a really concentrated solution.

3) Leave the solution <u>overnight</u> for the rest of the water to <u>evaporate</u>. Nice <u>big salt crystals</u> will form.

4) The reaction between <u>hydrochloric acid</u> and <u>sodium hydroxide</u> makes the salt <u>sodium chloride</u>:

$$\text{hydrochloric acid} + \text{sodium hydroxide} \longrightarrow \text{sodium chloride} + \text{water}$$

The Name of the Salt Depends on the Acid

1) <u>Hydrochloric acid</u> always reacts to make a salt with <u>chloride</u> in the <u>name</u>. For example, <u>sodium chloride</u>.

2) <u>Sulfuric acid</u> always reacts to make a salt with <u>sulfate</u> in the <u>name</u>. For example, <u>copper sulfate</u>.

It's fun making salts — but I wouldn't put them on chips...

Make sure you <u>know</u> which kinds of <u>salts</u> you get from which kind of <u>acid</u>. If you don't learn this stuff it'll come up at the <u>worst</u> possible opportunity. So you'd better get on with it...

Reactivity Series and Metal Extraction

You need to know which <u>metals</u> are <u>most reactive</u> — and which are <u>least reactive</u>.

The Reactivity Series — How Well a Metal Reacts

The <u>Reactivity Series</u> lists metals in <u>order</u> of how <u>reactive</u> they are.

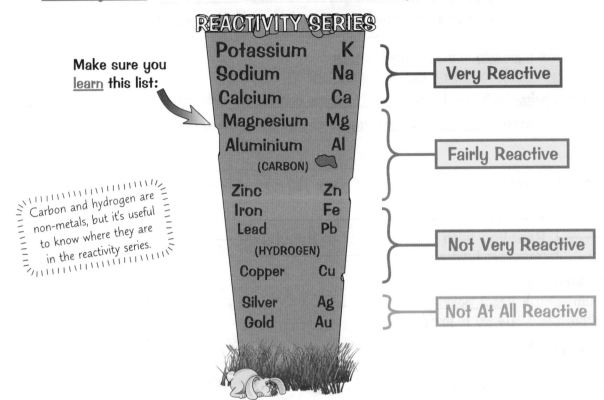

Make sure you <u>learn</u> this list:

Carbon and hydrogen are non-metals, but it's useful to know where they are in the reactivity series.

REACTIVITY SERIES

Potassium	K	Very Reactive
Sodium	Na	
Calcium	Ca	
Magnesium	Mg	Fairly Reactive
Aluminium	Al	
(CARBON)		
Zinc	Zn	Not Very Reactive
Iron	Fe	
Lead	Pb	
(HYDROGEN)		
Copper	Cu	
Silver	Ag	Not At All Reactive
Gold	Au	

Some Metals Can Be Extracted With Carbon

1) Metals are usually mined as <u>ores</u> — rocks containing <u>metal compounds</u>. These compounds are usually <u>metal oxides</u> — see page 53.

2) A metal can be <u>extracted</u> (removed) from its ore using <u>REDUCTION</u>.

3) Reduction means <u>removing oxygen</u> from the ore to leave behind the metal. Reduction of a metal ore can be done using <u>carbon</u>.

   For example, carbon is used to remove oxygen from iron oxide:

iron oxide + carbon ⟶ iron + carbon dioxide

4) Only metals <u>below</u> carbon in the reactivity series can be extracted from their ore using carbon.

5) Metals <u>above</u> carbon (like aluminium and magnesium) can't be extracted from their ores like this. You need <u>electricity</u> to get those out.

Potassium
Sodium
Calcium
Magnesium
Aluminium

—CARBON—

Zinc
Iron
Lead
Copper

Metal extraction — sounds painful...

This page contains some <u>important stuff</u> on metals. One of the first things you should know about metals is the difference in how <u>reactive</u> they are. In Chemistry, the <u>reactivity</u> of a metal is the most important feature of it, because that's what decides how it will <u>behave</u> in every reaction it's faced with.

Reaction of Metals with Acids

One more page on <u>metals</u> for you to learn — it's not so bad though, I promise. You don't need to know about each individual reaction, just how the <u>reactivity</u> of each metal affects it.

Reacting Metals With Dilute Acid

> metal + acid ⟶ salt + hydrogen

1) Metals above <u>hydrogen</u> in the <u>reactivity series</u> will <u>react</u> with <u>acids</u> to make a <u>salt</u> and <u>hydrogen</u>.

2) The metals <u>below</u> hydrogen in the <u>reactivity series don't react</u> with <u>acids</u>.

3) The reaction becomes <u>less and less exciting</u> as you go <u>down</u> the <u>series</u>.

More Reactive Metals React More Violently

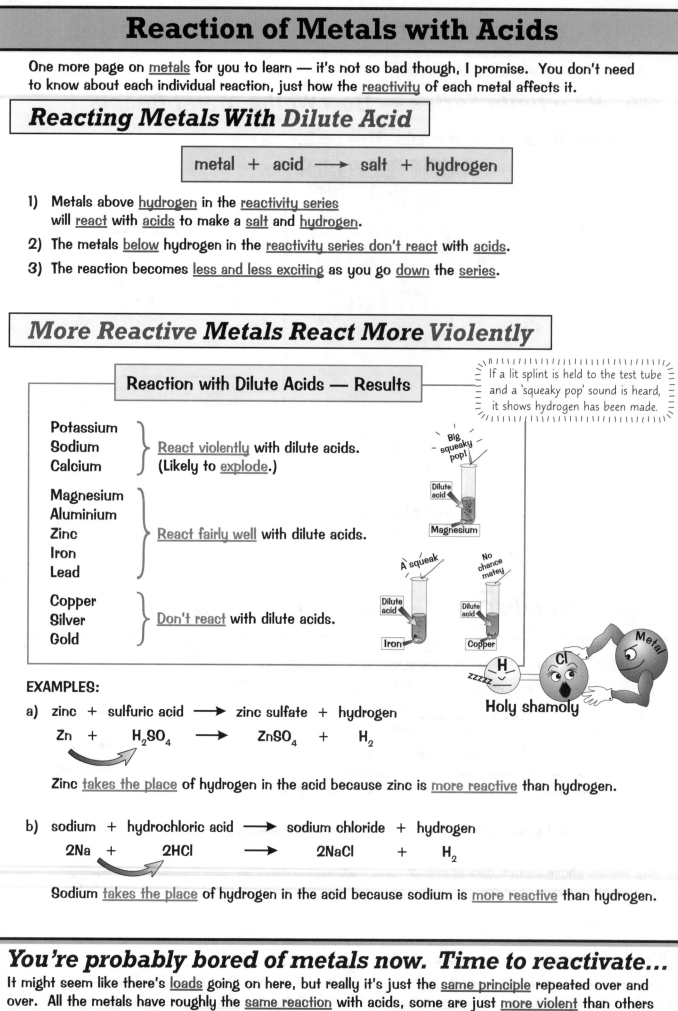

Reaction with Dilute Acids — Results

Potassium
Sodium   } <u>React violently</u> with dilute acids.
Calcium   (Likely to <u>explode</u>.)

Magnesium
Aluminium
Zinc   } <u>React fairly well</u> with dilute acids.
Iron
Lead

Copper
Silver   } <u>Don't react</u> with dilute acids.
Gold

If a lit splint is held to the test tube and a 'squeaky pop' sound is heard, it shows hydrogen has been made.

Big squeaky pop!

Dilute acid

Magnesium

A squeak

Dilute acid

Iron

No chance matey

Dilute acid

Copper

Holy shamoly

EXAMPLES:

a) zinc + sulfuric acid ⟶ zinc sulfate + hydrogen

Zn + H_2SO_4 ⟶ $ZnSO_4$ + H_2

Zinc <u>takes the place</u> of hydrogen in the acid because zinc is <u>more reactive</u> than hydrogen.

b) sodium + hydrochloric acid ⟶ sodium chloride + hydrogen

2Na + 2HCl ⟶ 2NaCl + H_2

Sodium <u>takes the place</u> of hydrogen in the acid because sodium is <u>more reactive</u> than hydrogen.

You're probably bored of metals now. Time to reactivate...

It might seem like there's <u>loads</u> going on here, but really it's just the <u>same principle</u> repeated over and over. All the metals have roughly the <u>same reaction</u> with acids, some are just <u>more violent</u> than others (and some don't happen at all). All you need to know is the <u>order</u> of how strongly they all react.

Reactions of Oxides with Acids

Oxides are exactly what they sound like — they've chemicals with oxygen in them somewhere...

Metals React With Oxygen to Make Oxides

Metals react with oxygen to make metal oxides.

Example: magnesium + oxygen → magnesium oxide

Metal Oxides are Alkaline

1) Metal oxides in solution have a pH which is higher than 7. They're alkaline.

2) So metal oxides react with acids to make a salt and water.

> acid + metal oxide ⟶ salt + water

EXAMPLES:
hydrochloric acid + copper oxide → copper chloride + water
sulfuric acid + zinc oxide → zinc sulfate + water

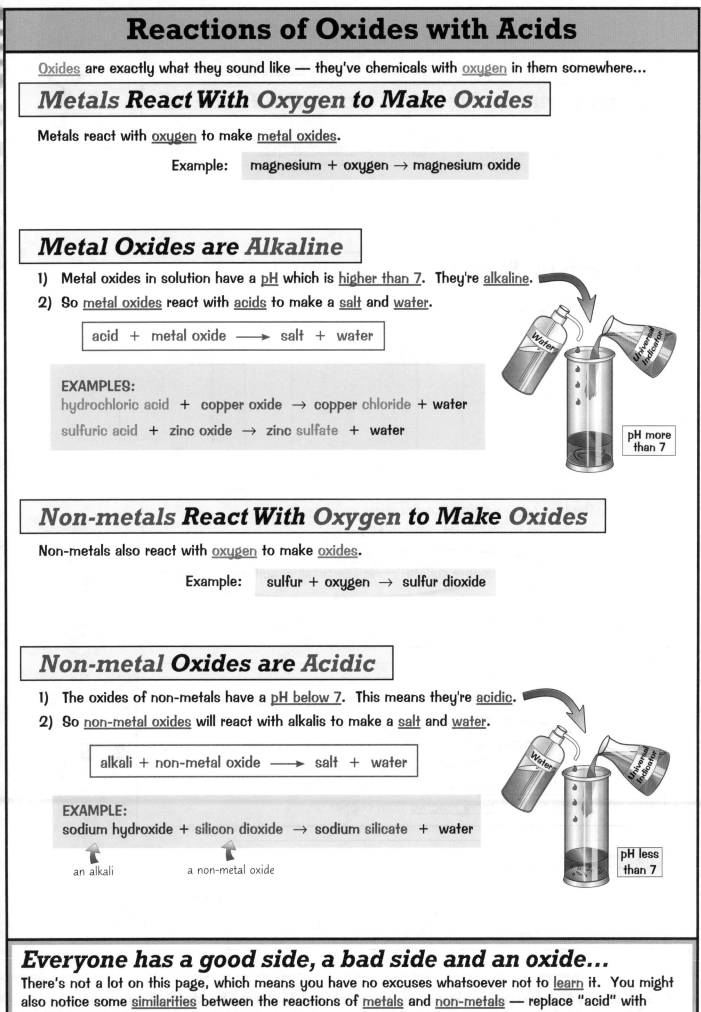

pH more than 7

Non-metals React With Oxygen to Make Oxides

Non-metals also react with oxygen to make oxides.

Example: sulfur + oxygen → sulfur dioxide

Non-metal Oxides are Acidic

1) The oxides of non-metals have a pH below 7. This means they're acidic.

2) So non-metal oxides will react with alkalis to make a salt and water.

> alkali + non-metal oxide ⟶ salt + water

EXAMPLE:
sodium hydroxide + silicon dioxide → sodium silicate + water

an alkali   a non-metal oxide

pH less than 7

Everyone has a good side, a bad side and an oxide...

There's not a lot on this page, which means you have no excuses whatsoever not to learn it. You might also notice some similarities between the reactions of metals and non-metals — replace "acid" with "alkali" and they're pretty much identical. With that said, you should try not to get the two confused.

Displacement Reactions

This page will make more sense if you can remember the <u>reactivity series</u> (page 51).

LEARN THIS RULE

'Displacement' Means 'Taking the Place of'

A <u>more reactive</u> metal will <u>displace</u> a <u>less reactive</u> metal from its compound.

1) In other words, a <u>more reactive</u> metal will <u>take the place</u> of a <u>less reactive</u> metal in a compound.

2) The <u>less reactive</u> metal gets <u>"kicked out"</u> of its compound.
It then <u>coats</u> itself on the reactive metal.

A Reactivity Series Investigation

You can use <u>displacement reactions</u> to <u>investigate</u> the <u>reactivity</u> of metals.

<u>Method:</u> 1) Slap a bit of metal into some salt solutions.   2) See what happens.

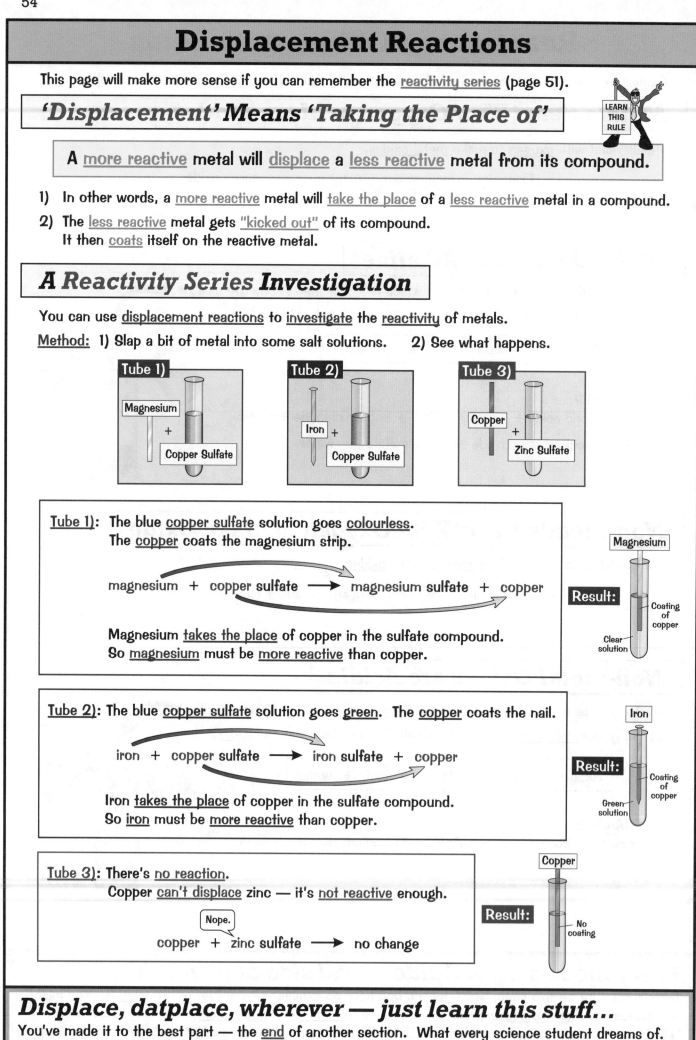

Tube 1) Magnesium + Copper Sulfate

Tube 2) Iron + Copper Sulfate

Tube 3) Copper + Zinc Sulfate

<u>Tube 1):</u> The blue <u>copper sulfate</u> solution goes <u>colourless</u>.
The <u>copper</u> coats the magnesium strip.

magnesium + copper sulfate ⟶ magnesium sulfate + copper

Magnesium <u>takes the place</u> of copper in the sulfate compound.
So <u>magnesium</u> must be <u>more reactive</u> than copper.

Magnesium
Result: Coating of copper — Clear solution

<u>Tube 2):</u> The blue <u>copper sulfate</u> solution goes <u>green</u>. The <u>copper</u> coats the nail.

iron + copper sulfate ⟶ iron sulfate + copper

Iron <u>takes the place</u> of copper in the sulfate compound.
So <u>iron</u> must be <u>more reactive</u> than copper.

Iron
Result: Coating of copper — Green solution

<u>Tube 3):</u> There's <u>no reaction</u>.
Copper <u>can't displace</u> zinc — it's <u>not reactive</u> enough.

Nope.

copper + zinc sulfate ⟶ no change

Copper
Result: No coating

Displace, datplace, wherever — just learn this stuff...

You've made it to the best part — the <u>end</u> of another section. What every science student dreams of.

Section Summary

Good work, you've made it through another section of Chemistry. Now it's time to sum it all up with a handy set of questions that test everything you need to know. Luckily for you, that's exactly what this page is for. You must have heard it all before by now — work through the questions one by one, make sure you know everything, then maybe treat yourself to something sweet.

1) In a chemical reaction, what are the chemicals you start with called?
   What are the chemicals you end up with called?
2) Name the two types of equation you can use to show what happens in a chemical reaction.
3) Which of the equations you named in question 2) shows how many atoms are on each side?
4) What's the chemical formula for a) water, b) table salt, c) carbon dioxide?
5) What happens to the atoms in a chemical reaction?
6) Does the mass change during a chemical reaction? Why or why not?
7) What's combustion?
8) A substance reacts with oxygen. What type of reaction is this?
9) What's thermal decomposition?
10) What's formed when copper carbonate breaks down by thermal decomposition?
11) What's the main difference between exothermic and endothermic reactions?
12)* If you put sodium in water, it catches fire and burns up.
    Is this reaction exothermic or endothermic?
13)* When ammonia breaks down to nitrogen and hydrogen, the temperature drops.
    Is this reaction exothermic or endothermic?
14) How does a catalyst affect the speed of a reaction?
15) How do catalysts make a reaction cheaper to run?
16) What pH does the strongest acid on a pH chart have? And the strongest alkali?
17) What pH does a neutral solution have?
18) What colour would universal indicator go if it was mixed with:
    a) a strong acid,   b) a neutral solution,   c) a strong alkali.
19) Complete this equation: acid + alkali ⟶ .... + ...
20) What type of reaction is shown in question 19)?
21) Hydrochloric acid makes chloride salts — what salts does sulfuric acid make?
22) What kind of salts do you get from nitric acid?
23) List the reactivity series in the correct order. Take the first letter of each element and make up a
    rhyme to help you remember it — there, that'll cheer you up.
24) If you haven't already, add carbon and hydrogen to your reactivity series from question 23).
25) Which metals in the reactivity series can be extracted from their ores using carbon?
    Which can't?
26) What do metals produce when they react with an acid?
27) Which metal will react the most violently with acid?
28) Are metal oxides in solution acidic, neutral or alkaline?
29) Give an example of a neutralisation reaction involving a metal oxide.
30) Are non-metal oxides in solution acidic, neutral or alkaline?
31) What does displacement mean?
32) What is the rule for displacement reactions?
33) Can magnesium displace copper from copper sulfate?

*Answers on page 108.

The Earth's Structure

Ever wondered what the planet's like on the <u>inside</u>? Well you're in for a <u>treat</u> with this page then.

The Earth Has a *Crust*, a *Mantle* and a *Core*

The Earth is <u>almost</u> a <u>sphere</u> and it has a <u>layered</u> structure. A bit like a scotch egg. Or a peach.

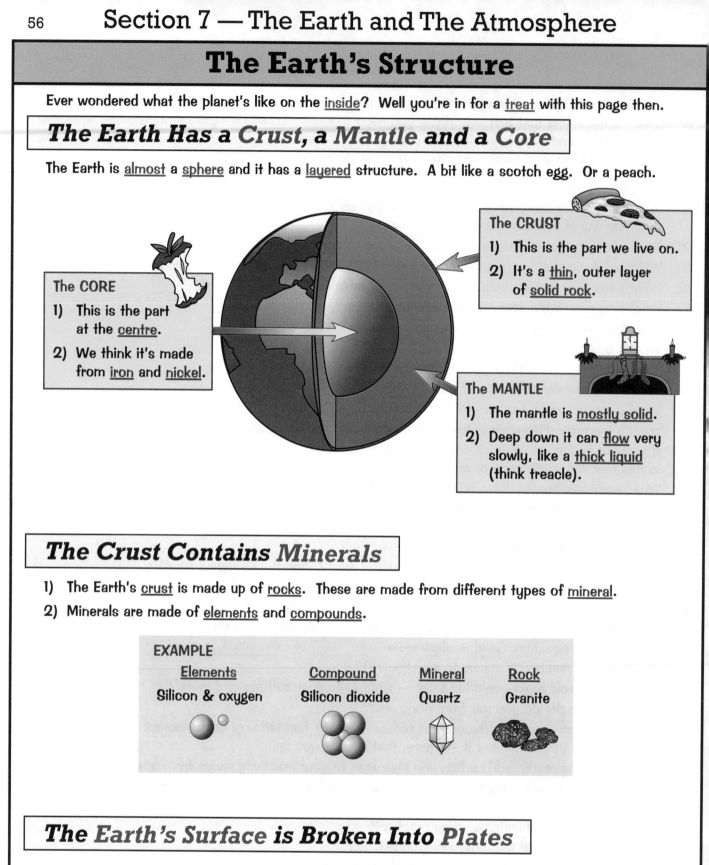

The CORE

1) This is the part at the <u>centre</u>.

2) We think it's made from <u>iron</u> and <u>nickel</u>.

The CRUST

1) This is the part we live on.

2) It's a <u>thin</u>, outer layer of <u>solid rock</u>.

The MANTLE

1) The mantle is <u>mostly solid</u>.

2) Deep down it can <u>flow</u> very slowly, like a <u>thick liquid</u> (think treacle).

The *Crust Contains* Minerals

1) The Earth's <u>crust</u> is made up of <u>rocks</u>. These are made from different types of <u>mineral</u>.

2) Minerals are made of <u>elements</u> and <u>compounds</u>.

EXAMPLE

Elements	Compound	Mineral	Rock
Silicon & oxygen	Silicon dioxide	Quartz	Granite

The *Earth's Surface* is Broken Into *Plates*

1) The crust and the upper part of the mantle are cracked into a number of <u>large plates</u>.

2) These plates are a bit like <u>big rafts</u> that 'float' on the mantle. They're able to <u>move</u> around slowly.

3) Sometimes, the plates move very <u>suddenly</u>, causing an <u>earthquake</u>.

Personally, I always cut off the crust...

You need to know the <u>structure</u> of <u>Earth</u>, i.e. what it would look like if you cut it open and what it's <u>made of</u>. That <u>top diagram</u> is your friend — <u>learn it</u> and learn it well. And, while we're on the subject, you'll need to learn all the <u>words</u> too. On the whole page. Phew.

Rock Types

Yep, there's <u>more than one</u> sort of rock. Who'd have thought it.

There are Three Different Types of Rock

1) Igneous Rocks

1) These are formed from <u>magma</u> (melted underground rock).

2) Some magma gets <u>pushed up</u> to the <u>surface</u> of the crust — and often out through <u>volcanoes</u>.

3) It then <u>cools</u> and forms rocks <u>above</u> ground.

4) Sometimes it cools <u>below</u> ground.

<u>EXAMPLES</u>: basalt (cooled above ground)
granite (cooled below ground)

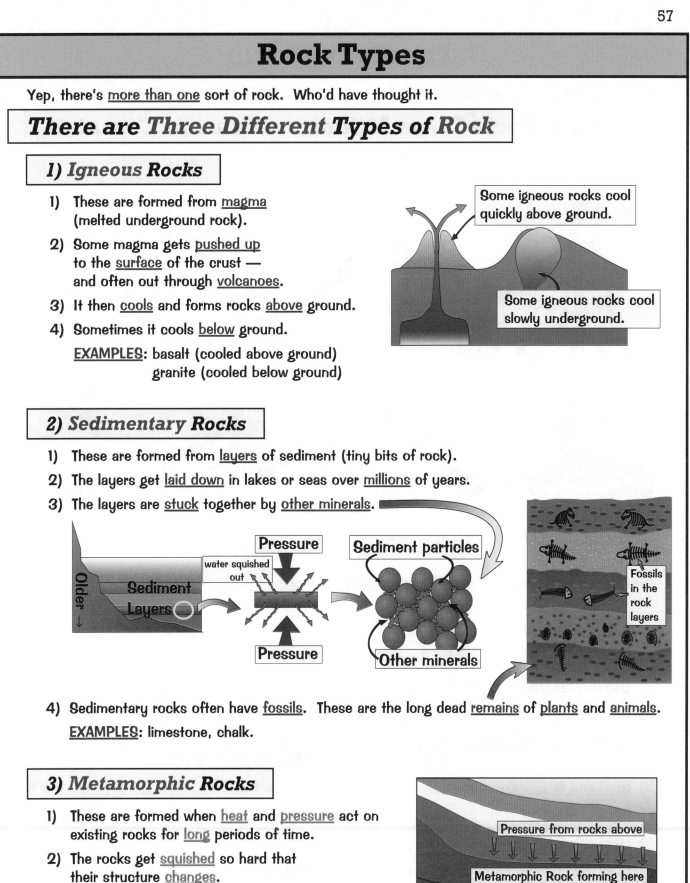

Some igneous rocks cool quickly above ground.

Some igneous rocks cool slowly underground.

2) Sedimentary Rocks

1) These are formed from <u>layers</u> of sediment (tiny bits of rock).

2) The layers get <u>laid down</u> in lakes or seas over <u>millions</u> of years.

3) The layers are <u>stuck</u> together by <u>other minerals</u>.

Pressure

water squished out

Older

Sediment Layers

Pressure

Sediment particles

Other minerals

Fossils in the rock layers

4) Sedimentary rocks often have <u>fossils</u>. These are the long dead <u>remains</u> of <u>plants</u> and <u>animals</u>.
<u>EXAMPLES</u>: limestone, chalk.

3) Metamorphic Rocks

1) These are formed when <u>heat</u> and <u>pressure</u> act on existing rocks for <u>long</u> periods of time.

2) The rocks get <u>squished</u> so hard that their structure <u>changes</u>.

3) Metamorphic rocks may have really <u>tiny crystals</u>. Some have layers.

<u>EXAMPLES</u>: marble, slate.

Pressure from rocks above

Metamorphic Rock forming here

Magma   Intense heat from below

OK, sure — but doesn't pop rock deserve a mention...

Three types of rock for you to know about — make sure you do. I agree they sound pretty scary, but once you get over that little hurdle and just <u>learn them</u>, the rest follows a lot easier.

The Rock Cycle

The rock cycle involves changes to rocks both <u>inside</u> and <u>outside</u> the Earth.

The Rock Cycle Takes Millions of Years to Complete

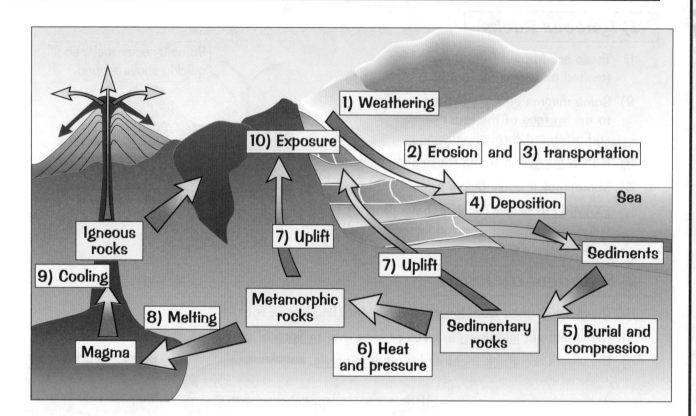

The three types of rock are <u>igneous</u>, <u>sedimentary</u> and <u>metamorphic</u> (see page 57).
The rock cycle involves <u>changing</u> the three types of rock from one to another. This happens by:

1) <u>WEATHERING</u>: <u>breaking down</u> rocks into <u>smaller bits</u>.

2) <u>EROSION</u>: <u>wearing down</u> rocks, for example, by rain.

3) <u>TRANSPORTATION</u>: <u>moving</u> the eroded bits of rock round the world by <u>wind</u> and <u>water</u> (mostly).

4) <u>DEPOSITION</u>: laying down of <u>sediment</u>.

5) <u>BURIAL and COMPRESSION</u>: <u>squeezing</u> and <u>compressing</u> the layers —
                 eventually they form <u>SEDIMENTARY ROCKS</u>.

6) <u>HEAT and PRESSURE</u>: further <u>squashing</u> and <u>heating</u> turns the rocks into <u>METAMORPHIC ROCKS</u>.

7) <u>UPLIFT</u>: rocks are <u>pushed up</u> to the surface.

8) <u>MELTING</u>: lots of <u>heat</u> makes the rocks <u>melt</u> a little — that changes them to magma.

9) <u>COOLING</u>: The molten (melted) rock turns to <u>solid</u> <u>IGNEOUS ROCK</u>.

10) <u>EXPOSURE</u>: <u>back</u> to weathering and erosion again.

The Rock Cycle's like homework — it takes forever...

<u>Ten stages</u> of the rock cycle to learn there — make sure you know <u>what happens</u> at each stage, but also how each of the stages are <u>linked</u>. You'll really impress teacher if you can <u>explain</u> how a <u>sedimentary</u> <u>rock</u> changes into a <u>metamorphic rock</u>, and so on. Why not try writing a <u>summary</u> in your own words?

Recycling

My mum is pretty big on recycling. You can't throw anything in the bin in our house...

The Earth is the Source of All Our Resources

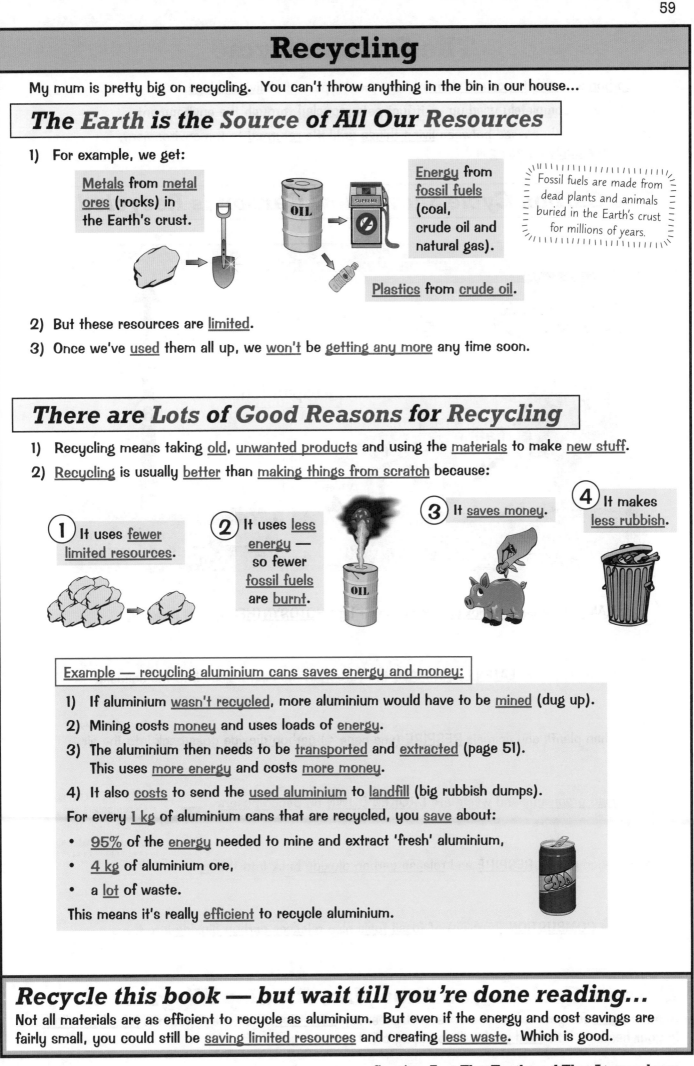

1) For example, we get:

Metals from **metal ores** (rocks) in the Earth's crust.

Energy from **fossil fuels** (coal, crude oil and natural gas).

Fossil fuels are made from dead plants and animals buried in the Earth's crust for millions of years.

Plastics from **crude oil**.

2) But these resources are **limited**.

3) Once we've **used** them all up, we **won't** be **getting any more** any time soon.

There are Lots of Good Reasons for Recycling

1) Recycling means taking **old**, **unwanted products** and using the **materials** to make **new stuff**.

2) **Recycling** is usually **better** than **making things from scratch** because:

① It uses **fewer limited resources**.

② It uses **less energy** — so fewer **fossil fuels** are **burnt**.

③ It **saves money**.

④ It makes **less rubbish**.

Example — recycling aluminium cans saves energy and money:

1) If aluminium **wasn't recycled**, more aluminium would have to be **mined** (dug up).

2) Mining costs **money** and uses loads of **energy**.

3) The aluminium then needs to be **transported** and **extracted** (page 51). This uses **more energy** and costs **more money**.

4) It also **costs** to send the **used aluminium** to **landfill** (big rubbish dumps).

For every **1 kg** of aluminium cans that are recycled, you **save** about:

- **95%** of the **energy** needed to mine and extract 'fresh' aluminium,
- **4 kg** of aluminium ore,
- a **lot** of waste.

This means it's really **efficient** to recycle aluminium.

Recycle this book — but wait till you're done reading...

Not all materials are as efficient to recycle as aluminium. But even if the energy and cost savings are fairly small, you could still be **saving limited resources** and creating **less waste**. Which is good.

The Carbon Cycle

1) <u>Carbon</u> is a <u>very important element</u> because it's part of all <u>living things</u>.

2) It's never completely <u>used up</u> — it just gets <u>recycled</u> through the environment.

3) It's constantly passed between <u>living things</u> until it's <u>returned</u> to the environment, and the <u>cycle</u> starts again.

The Carbon Cycle Shows How Carbon is Recycled

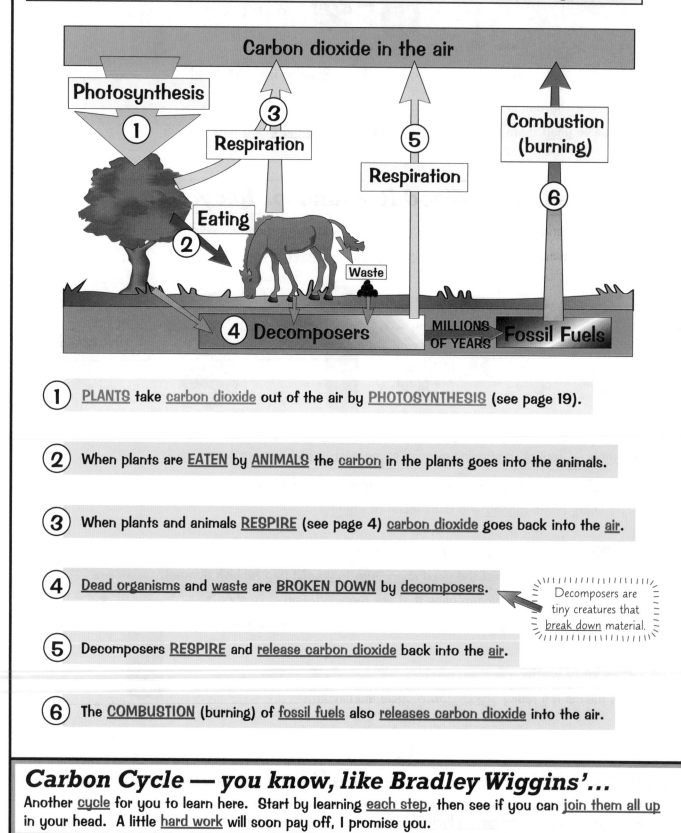

① <u>PLANTS</u> take <u>carbon dioxide</u> out of the air by <u>PHOTOSYNTHESIS</u> (see page 19).

② When plants are <u>EATEN</u> by <u>ANIMALS</u> the <u>carbon</u> in the plants goes into the animals.

③ When plants and animals <u>RESPIRE</u> (see page 4) <u>carbon dioxide</u> goes back into the <u>air</u>.

④ <u>Dead organisms</u> and <u>waste</u> are <u>BROKEN DOWN</u> by <u>decomposers</u>.

Decomposers are tiny creatures that <u>break down</u> material.

⑤ Decomposers <u>RESPIRE</u> and <u>release carbon dioxide</u> back into the <u>air</u>.

⑥ The <u>COMBUSTION</u> (burning) of <u>fossil fuels</u> also <u>releases carbon dioxide</u> into the air.

Carbon Cycle — you know, like Bradley Wiggins'...

Another <u>cycle</u> for you to learn here. Start by learning <u>each step</u>, then see if you can <u>join them all up</u> in your head. A little <u>hard work</u> will soon pay off, I promise you.

The Atmosphere and Climate

It's important to know exactly what you're <u>breathing</u> in and out. So read this page and find out.

The Earth's Atmosphere is Made Up of Different Gases

1) The <u>gases</u> that surround a planet make up that planet's <u>atmosphere</u>.

2) The <u>Earth's atmosphere</u> is around:

78% nitrogen (N_2)   **21% oxygen (O_2)**   **0.04% carbon dioxide (CO_2)**

3) It also contains <u>small amounts</u> of other gases, like <u>water vapour</u>.
(There's <u>more</u> water vapour than carbon dioxide in the atmosphere.)

The Carbon Dioxide Level is Increasing

The level of carbon dioxide in the Earth's <u>atmosphere</u> is rising — and it's down to <u>human activities</u> and some <u>natural causes</u>. Here are some <u>examples</u> of human activities that affect carbon dioxide levels:

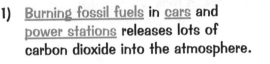

1) <u>Burning fossil fuels</u> in <u>cars</u> and <u>power stations</u> releases lots of carbon dioxide into the atmosphere.

2) <u>Deforestation</u> (chopping down trees) means <u>less carbon dioxide</u> is <u>removed</u> from the atmosphere by <u>photosynthesis</u>.

Carbon Dioxide Affects the Earth's Climate

1) Carbon dioxide <u>traps energy</u> from the <u>Sun</u> in the <u>Earth's atmosphere</u>.

2) This <u>stops</u> some energy from <u>being lost</u> into space and helps to keep the <u>Earth warm</u>.

3) But the <u>level</u> of <u>carbon dioxide</u> is <u>increasing</u>.

4) And the <u>Earth</u> has been getting <u>hotter</u>. Most scientists think this is <u>because of</u> the rise in <u>carbon dioxide</u> levels.

5) This increase in the Earth's temperature is called <u>global warming</u>.

6) Global warming could have some <u>serious effects</u>. For example:

- <u>Ice</u> on land (like the large <u>sheets of ice</u> covering Greenland) might melt faster. This could cause <u>sea levels</u> to <u>rise</u>, leading to <u>floods</u>.

- <u>Rainfall patterns</u> could change. This might make it <u>harder</u> for farmers to <u>grow crops</u>.

Is it just me or is it getting really warm in here...?

You need to know what makes up the Earth's <u>atmosphere</u> — so get learning those <u>percentages</u> at the top of the page. You also need to <u>understand</u> how <u>human activities</u> are affecting the <u>Earth's climate</u>.

Section Summary

Well there we are. The end of Section 7. All you have to do now is learn it all. And yes you've guessed it, here below are some lovely questions I prepared earlier. It's no good going through them and only answering the one or two that take your fancy. Make sure you can answer all of them.

1) The Earth is covered with a thin outer layer of rock. What is this layer called?

2) What is the name of the structure between the outer layer of rock and the Earth's core? Explain why it can flow near the bottom even though it's solid at the top.

3) Which two metals do we think the Earth's core is made of?

4) Name a mineral present in the Earth's crust.

5) What's the Earth's crust broken into?

6) How are igneous rocks formed?

7) How do sedimentary rocks form?

8) The dead remains of plants and animals can become trapped in sedimentary rocks. What are these remains called?

9) How do metamorphic rocks form?

10) Give two examples of: a) igneous rocks, b) sedimentary rocks, c) metamorphic rocks.

11) In the rock cycle, describe what happens during:
   a) erosion   b) deposition   c) uplift.

12) What must happen to sedimentary rocks to turn them into metamorphic rocks?

13) What must happen to metamorphic rocks to turn them into igneous rocks?

14) Name two limited resources we get from the Earth.

15) Give four reasons why it's good to recycle materials.

16) Which living things remove carbon dioxide from the air?

17) How does carbon get passed from plants to animals?

18) a) How do plants, animals and decomposers all return carbon dioxide to the air?
   b) How else is carbon dioxide returned to the air?

19) What percentage of the Earth's atmosphere is: a) nitrogen, b) oxygen, c) carbon dioxide?

20) Name one other gas present in the Earth's atmosphere.

21) Give two human activities that are increasing the level of carbon dioxide in the atmosphere. Say why each one has an effect on the level of CO_2.

22) How does carbon dioxide help to keep the Earth warm?

23) What is global warming? What's causing it?

24) Describe two possible effects of global warming.

Energy Stores

Energy is everywhere, so this is a pretty important page.

Learn These Seven Energy Stores...

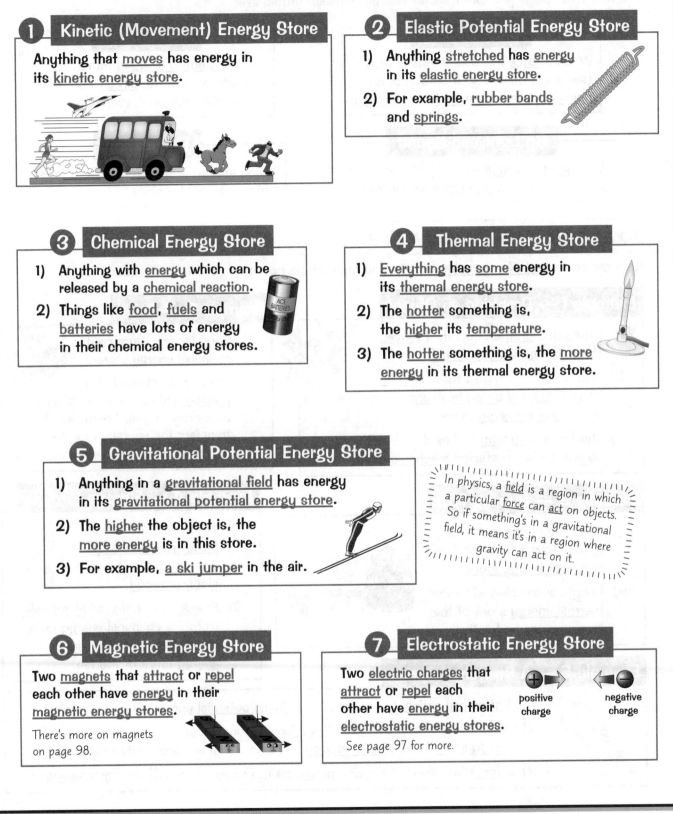

1 Kinetic (Movement) Energy Store

Anything that <u>moves</u> has energy in its <u>kinetic energy store</u>.

2 Elastic Potential Energy Store

1) Anything <u>stretched</u> has <u>energy</u> in its <u>elastic energy store</u>.
2) For example, <u>rubber bands</u> and <u>springs</u>.

3 Chemical Energy Store

1) Anything with <u>energy</u> which can be released by a <u>chemical reaction</u>.
2) Things like <u>food</u>, <u>fuels</u> and <u>batteries</u> have lots of energy in their chemical energy stores.

ACE BATTERIES

4 Thermal Energy Store

1) <u>Everything</u> has <u>some</u> energy in its <u>thermal energy store</u>.
2) The <u>hotter</u> something is, the <u>higher</u> its <u>temperature</u>.
3) The <u>hotter</u> something is, the <u>more</u> <u>energy</u> in its thermal energy store.

5 Gravitational Potential Energy Store

1) Anything in a <u>gravitational field</u> has energy in its <u>gravitational potential energy store</u>.
2) The <u>higher</u> the object is, the <u>more energy</u> is in this store.
3) For example, <u>a ski jumper</u> in the air.

In physics, a <u>field</u> is a region in which a particular <u>force</u> can <u>act</u> on objects. So if something's in a gravitational field, it means it's in a region where gravity can act on it.

6 Magnetic Energy Store

Two <u>magnets</u> that <u>attract</u> or <u>repel</u> each other have <u>energy</u> in their <u>magnetic energy stores</u>.

There's more on magnets on page 98.

7 Electrostatic Energy Store

Two <u>electric charges</u> that <u>attract</u> or <u>repel</u> each other have <u>energy</u> in their <u>electrostatic energy stores</u>.

See page 97 for more.

positive charge   negative charge

Start of a new section — I'm feeling energised...

Seven energy stores for you to get your head around — you need to make sure you learn <u>all of them</u>. Grab a piece of <u>paper</u>, <u>cover up</u> the page and <u>scribble down</u> as much as you can. See if you can name and describe all seven. If you can't, keep trying until you can. You'll be an energy expert in no time.

Energy Transfer

And now it's time to see how <u>energy</u> can be <u>moved between stores</u>...

There are Four Ways of Transferring Energy

The four main ways you can transfer energy between stores are:

Mechanically

When a <u>force</u> makes something <u>move</u> (see page 65). E.g. if an object is <u>pushed</u>, <u>pulled</u>, <u>stretched</u> or <u>squashed</u>.

Electrically

When <u>electric charges</u> move around an electric <u>circuit</u> due to a potential difference (see page 94).

By Light and Sound

When <u>light</u> or <u>sound</u> waves (see Section 10) carry energy from <u>one place</u> to <u>another</u>.

By Heating

When energy is transferred from <u>hotter</u> objects to <u>colder</u> objects (see page 66).

Examples of Energy Transfer

Here are some <u>real-life examples</u> to help you understand how energy is transferred.

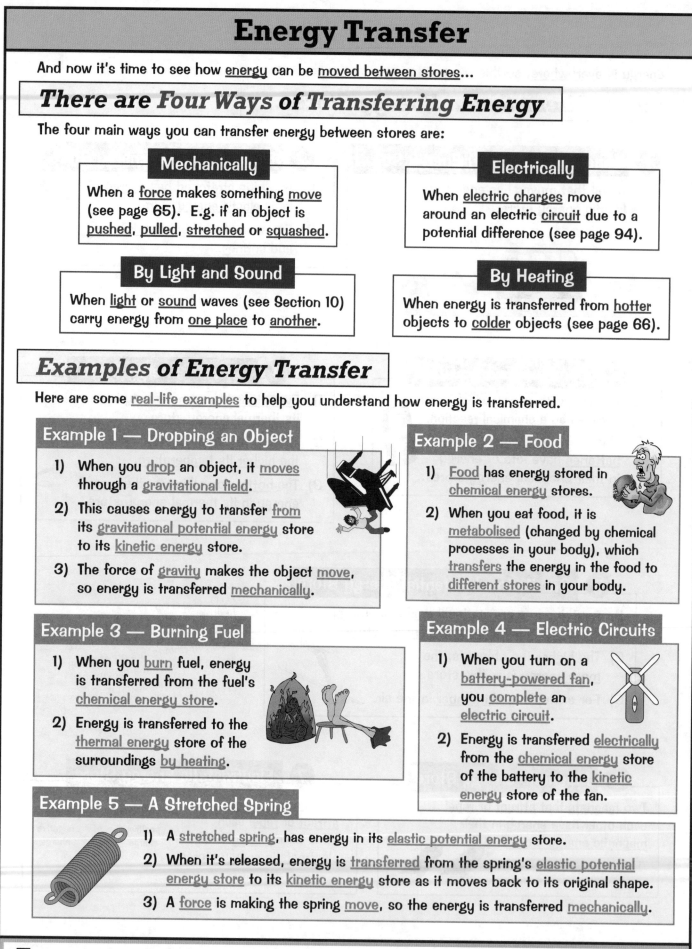

Example 1 — Dropping an Object

1) When you <u>drop</u> an object, it <u>moves</u> through a <u>gravitational field</u>.

2) This causes energy to transfer <u>from</u> its <u>gravitational potential energy</u> store to its <u>kinetic energy</u> store.

3) The force of <u>gravity</u> makes the object <u>move</u>, so energy is transferred <u>mechanically</u>.

Example 2 — Food

1) <u>Food</u> has energy stored in <u>chemical energy</u> stores.

2) When you eat food, it is <u>metabolised</u> (changed by chemical processes in your body), which <u>transfers</u> the energy in the food to <u>different stores</u> in your body.

Example 3 — Burning Fuel

1) When you <u>burn</u> fuel, energy is transferred from the fuel's <u>chemical energy store</u>.

2) Energy is transferred to the <u>thermal energy</u> store of the surroundings <u>by heating</u>.

Example 4 — Electric Circuits

1) When you turn on a <u>battery-powered fan</u>, you <u>complete</u> an <u>electric circuit</u>.

2) Energy is transferred <u>electrically</u> from the <u>chemical energy</u> store of the battery to the <u>kinetic energy</u> store of the fan.

Example 5 — A Stretched Spring

1) A <u>stretched spring</u>, has energy in its <u>elastic potential energy</u> store.

2) When it's released, energy is <u>transferred</u> from the spring's <u>elastic potential energy store</u> to its <u>kinetic energy</u> store as it moves back to its original shape.

3) A <u>force</u> is making the spring <u>move</u>, so the energy is transferred <u>mechanically</u>.

Energy transfers — a lot cheaper than Man United's...

Well, I never knew <u>energy transfers</u> could be so much fun. Make sure you know the <u>four ways listed</u>. Then get your head round those <u>real-life examples</u>. Next transfer some energy to the kinetic energy store of your pen and try writing out the <u>whole page</u> to really get it stuck in your mind.

More Energy Transfer

Now it's time for more on <u>mechanical energy transfers</u> — when a <u>force</u> makes an object <u>move</u>...

Energy is Transferred When a Force Moves an Object

> When a <u>force moves</u> an object through a <u>distance</u>, <u>energy is transferred</u>.

Energy transferred is the same as work done — see page 80.

1) Objects need a <u>force</u> to make them <u>move</u>.

2) <u>Energy</u> is needed to <u>supply</u> this force.

3) The energy supplied is <u>transferred</u> to the object's <u>kinetic energy store</u>, so the object moves.

Example — Pushing a Broom

1) Whenever something <u>moves</u>, something else is supplying some sort of '<u>effort</u>' to move it.

2) The thing putting in the <u>effort</u> needs a <u>supply</u> of <u>energy</u> (from <u>fuel</u> or <u>food</u>, etc.).

3) It then <u>transfers energy</u> by <u>moving</u> the object — the supply of energy is transferred to <u>kinetic energy stores</u>.

Energy supplied

Energy transferred

One More Thing About Energy Transfer and Forces...

If you give a machine a <u>set amount</u> of energy to transfer, it can either:

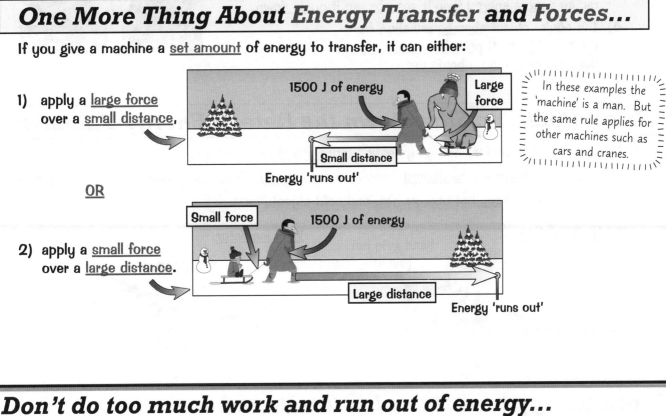

1) apply a <u>large force</u> over a <u>small distance</u>,

1500 J of energy — Large force — Small distance — Energy 'runs out'

In these examples the 'machine' is a man. But the same rule applies for other machines such as cars and cranes.

OR

2) apply a <u>small force</u> over a <u>large distance</u>.

Small force — 1500 J of energy — Large distance — Energy 'runs out'

Don't do too much work and run out of energy...

Make sure you understand how <u>forces</u> do work and cause <u>energy transfers</u>. Cover up the page, scribble down what you can remember then see what you've missed. Then feel free to laugh at that <u>lazy elephant</u>.

Energy Transfer by Heating

Energy *is* Transferred *From Hotter to Cooler Objects*

1) When two objects are at <u>different temperatures</u>, the <u>hotter</u> object transfers <u>energy</u> to the <u>cooler</u> object.

2) This carries on until both objects reach the <u>same temperature</u>. At this point we say they've reached <u>THERMAL EQUILIBRIUM</u>.

You need to know about <u>two ways</u> in which energy can be transferred between objects by <u>heating</u>:

1) Conduction

1) When an object is <u>heated</u>, its particles start <u>vibrating</u> (shaking) more. The particles have <u>more energy</u>.

2) Vibrating particles <u>transfer</u> energy when they <u>bump into</u> other particles that aren't vibrating as much.

3) When a <u>hot</u> and <u>cold</u> object are <u>touching</u>, particles in the hot object <u>transfer energy</u> to particles in the cold object.

4) The hot object <u>loses energy</u> so it <u>cools down</u>. The cold object <u>gains energy</u> so it <u>heats up</u>.

a particle

Energy transfer

Hot object   Cold object

2) Radiation

1) <u>All objects</u> radiate (send out) invisible <u>waves</u> (called <u>radiation</u>) to the surroundings. These waves can <u>transfer energy</u> from one place to another.

2) Objects can also <u>absorb</u> (take in) radiation.

3) The hotter object (like this hot potato) <u>radiates more energy than it absorbs</u>, so it <u>cools down</u>.

4) The cooler object <u>absorbs</u> radiation from the hot object. It <u>absorbs more energy than it radiates</u>, so it <u>heats up</u>.

Radiation

Hot potato   Cool potato

Insulators *Can Slow Down the Rate of Energy Transfer*

1) Some materials transfer energy by heating <u>more quickly</u> than others.

2) Materials like <u>plastic</u> and <u>cardboard</u> transfer energy <u>slowly</u>. These materials are called <u>INSULATORS</u>.

3) Insulators help to <u>keep</u> hot objects hot, and cold objects cold.

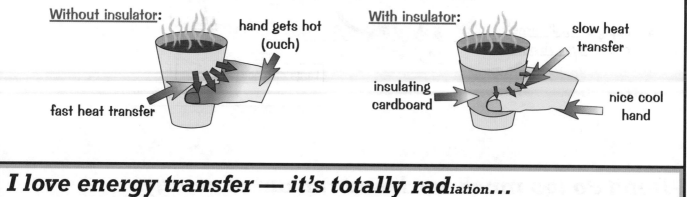

<u>Without insulator:</u>

hand gets hot (ouch)

fast heat transfer

<u>With insulator:</u>

slow heat transfer

insulating cardboard

nice cool hand

*I love energy transfer — it's totally rad*iation...

Make sure you learn the differences between the <u>two methods</u> of energy transfer on this page. One big difference is that <u>conduction</u> needs two objects to be <u>touching</u>, but <u>radiation doesn't</u>.

Conservation of Energy

Energy <u>has to</u> be transferred. Not always in a <u>useful</u> way though.

Some Laws About Energy

You need to know two <u>important laws</u> relating to energy:

1

> Energy can never be CREATED nor DESTROYED
> — it's only ever TRANSFERRED from one store to another.

This means energy is <u>conserved</u> (it never disappears).

2

> Energy is ONLY USEFUL when it's TRANSFERRED from one store to another.

Most Energy Transfers are Not Perfect

1) Useful devices <u>transfer energy</u> from <u>one store</u> to <u>another</u>.

2) Some of the <u>energy</u> put into the device will be transferred <u>usefully</u>. Some will be <u>wasted</u> — usually <u>by heating</u>.

3) BUT <u>no energy is destroyed</u>:

> Energy INPUT = USEFUL Energy + WASTED Energy

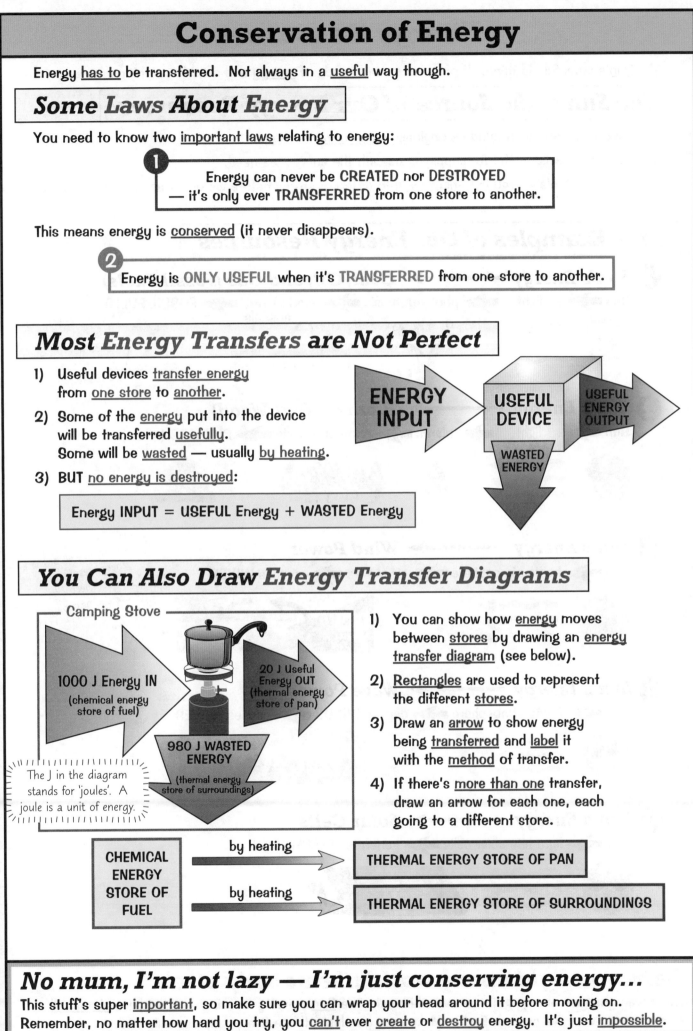

ENERGY INPUT → USEFUL DEVICE → USEFUL ENERGY OUTPUT → WASTED ENERGY

You Can Also Draw Energy Transfer Diagrams

Camping Stove

1000 J Energy IN (chemical energy store of fuel)

20 J Useful Energy OUT (thermal energy store of pan)

980 J WASTED ENERGY (thermal energy store of surroundings)

The J in the diagram stands for 'joules'. A joule is a unit of energy.

1) You can show how <u>energy</u> moves between <u>stores</u> by drawing an <u>energy transfer diagram</u> (see below).

2) <u>Rectangles</u> are used to represent the different <u>stores</u>.

3) Draw an <u>arrow</u> to show energy being <u>transferred</u> and <u>label</u> it with the <u>method</u> of transfer.

4) If there's <u>more than one</u> transfer, draw an arrow for each one, each going to a different store.

CHEMICAL ENERGY STORE OF FUEL — by heating → THERMAL ENERGY STORE OF PAN

— by heating → THERMAL ENERGY STORE OF SURROUNDINGS

No mum, I'm not lazy — I'm just conserving energy...

This stuff's super <u>important</u>, so make sure you can wrap your head around it before moving on. Remember, no matter how hard you try, you <u>can't</u> ever <u>create</u> or <u>destroy</u> energy. It's just <u>impossible</u>.

Energy Resources

The <u>Sun</u>'s a useful old thing. It provides us with loads of <u>energy</u> and asks for nothing in return.

The Sun is the Source of Our Energy Resources

1) Most of the <u>energy</u> around us originally <u>came from</u> the <u>Sun</u>.

2) The Sun is really useful for supplying us with the energy we need.

3) Often the Sun's energy is <u>transferred</u> into <u>different stores</u> before we use it as an <u>energy resource</u>.

Five Examples of Our Energy Resources

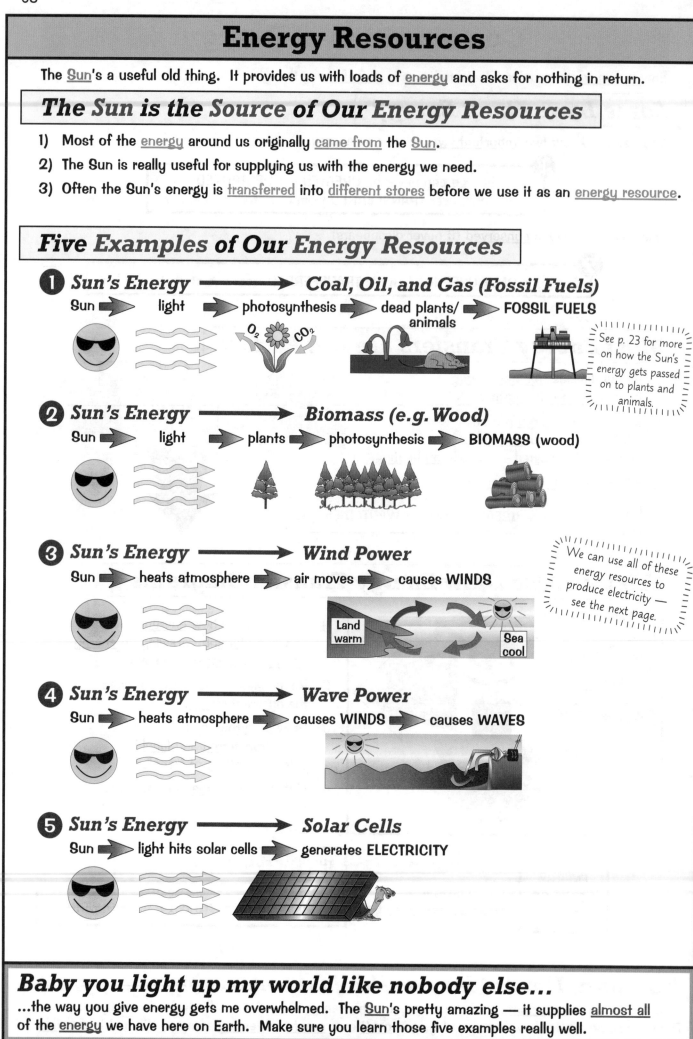

1 *Sun's Energy* ⟶ *Coal, Oil, and Gas (Fossil Fuels)*

Sun ➡ light ➡ photosynthesis ➡ dead plants/animals ➡ FOSSIL FUELS

O_2 CO_2

See p. 23 for more on how the Sun's energy gets passed on to plants and animals.

2 *Sun's Energy* ⟶ *Biomass (e.g. Wood)*

Sun ➡ light ➡ plants ➡ photosynthesis ➡ BIOMASS (wood)

3 *Sun's Energy* ⟶ *Wind Power*

Sun ➡ heats atmosphere ➡ air moves ➡ causes WINDS

Land warm   Sea cool

We can use all of these energy resources to produce electricity — see the next page.

4 *Sun's Energy* ⟶ *Wave Power*

Sun ➡ heats atmosphere ➡ causes WINDS ➡ causes WAVES

5 *Sun's Energy* ⟶ *Solar Cells*

Sun ➡ light hits solar cells ➡ generates ELECTRICITY

Baby you light up my world like nobody else...

...the way you give energy gets me overwhelmed. The <u>Sun</u>'s pretty amazing — it supplies <u>almost all</u> of the <u>energy</u> we have here on Earth. Make sure you learn those five examples really well.

Generating Electricity

We can use the <u>energy</u> we get from the Sun to <u>generate electricity</u>, in lots of different ways...

There Are Different Ways of Generating Electricity

1) We can use <u>energy resources</u> to <u>generate</u> (make) <u>electricity</u>.

2) At the moment we generate most of our electricity by burning <u>fossil fuels</u>.

3) Most ways of <u>generating electricity</u> turn a <u>turbine</u> and a <u>generator</u>. The generator transfers energy from <u>kinetic energy stores</u> away <u>electrically</u>.

> See previous page for more on energy resources.

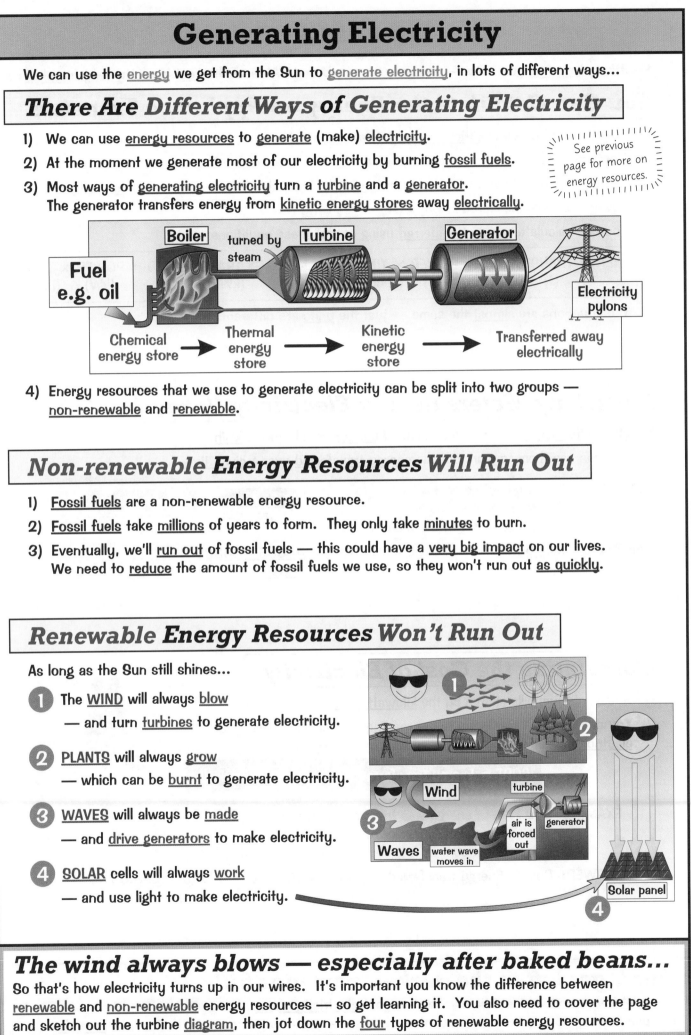

4) Energy resources that we use to generate electricity can be split into two groups — <u>non-renewable</u> and <u>renewable</u>.

Non-renewable Energy Resources Will Run Out

1) <u>Fossil fuels</u> are a non-renewable energy resource.

2) <u>Fossil fuels</u> take <u>millions</u> of years to form. They only take <u>minutes</u> to burn.

3) Eventually, we'll <u>run out</u> of fossil fuels — this could have a <u>very big impact</u> on our lives. We need to <u>reduce</u> the amount of fossil fuels we use, so they won't run out <u>as quickly</u>.

Renewable Energy Resources Won't Run Out

As long as the Sun still shines...

1) The <u>WIND</u> will always <u>blow</u>
   — and turn <u>turbines</u> to generate electricity.

2) <u>PLANTS</u> will always <u>grow</u>
   — which can be <u>burnt</u> to generate electricity.

3) <u>WAVES</u> will always be <u>made</u>
   — and <u>drive generators</u> to make electricity.

4) <u>SOLAR</u> cells will always <u>work</u>
   — and use light to make electricity.

The wind always blows — especially after baked beans...

So that's how electricity turns up in our wires. It's important you know the difference between <u>renewable</u> and <u>non-renewable</u> energy resources — so get learning it. You also need to cover the page and sketch out the turbine <u>diagram</u>, then jot down the <u>four</u> types of renewable energy resources.

The Cost of Electricity

Electricity isn't free you know — ask your mum and dad. At least the cost is pretty easy to calculate.

You Can Calculate the Energy an Appliance Transfers

1) Anything that needs electricity to work is an electrical appliance.

2) Electrical appliances transfer energy between stores (see page 64).

3) Energy transferred can be measured in joules (J), kilojoules (kJ) or kilowatt-hours (kWh).

4) Power is usually measured in watts (W) or kilowatts (kW).

5) You can calculate energy transferred using one of these equations:

> ENERGY TRANSFERRED = POWER × TIME
> (J)   (W)   (seconds)

> ENERGY TRANSFERRED = POWER × TIME
> (kWh)   (kW)   (hours)

The equations are almost the same — just the units are different.

Electricity Meters Record Electricity Usage

1) Electricity meters record the amount of energy transferred in kWh.

2) You can use them to work out the energy transferred over a given time.

> **EXAMPLE:** Here are two meter readings. How much energy has been transferred between 6 pm and 6 am?
>
> 4 4 3 8 0 kWh → 4 4 3 8 5 kWh
> 6 pm   6 am
>
> **ANSWER:** Energy transferred from 6 pm to 6 am = 44385 − 44380
> = 5 kWh
>
> You need to work out the difference between the two meter readings.

Calculating the Cost of Electricity

1) Household fuel bills charge by the kilowatt-hour.

2) You can use your electricity meter readings to calculate what your electricity bill should be. You need to use this formula:

> **COST = Energy transferred (kWh) × PRICE per kWh**

> **EXAMPLE:** 350 kWh of energy was transferred by electricity to Jo's home last month. Electricity costs 16p per kWh.
> Calculate the cost of Jo's electricity bill last month.
>
> **ANSWER:** Cost = Energy transferred × price = 350 × 16 = 5600p = £56.00

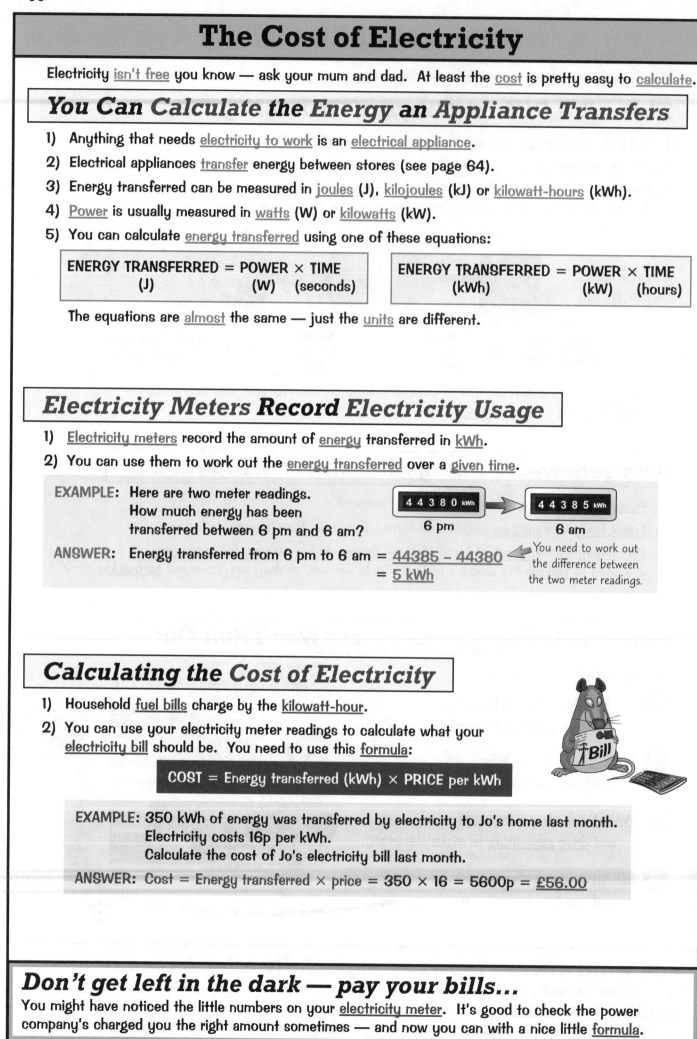

Don't get left in the dark — pay your bills...

You might have noticed the little numbers on your electricity meter. It's good to check the power company's charged you the right amount sometimes — and now you can with a nice little formula.

Comparing Power Ratings and Energy Values

You can work out how much energy an <u>appliance</u> transfers if you know its <u>power rating</u>. And to find out how much energy is in your <u>food</u>, just check the label. This energy stuff's everywhere.

Power Ratings of Appliances

1) The <u>power rating</u> of an appliance tells you how <u>fast</u> it <u>transfers energy</u>.

2) You can use an appliance's power rating to <u>work out</u> the energy it transfers in a certain <u>time</u>.

3) To do this you need to use the <u>equations</u> on the <u>previous page</u>.

The Energy Transferred Depends on the Power Rating

1) The higher the <u>power rating</u> of an appliance, the <u>more energy</u> it transfers in a <u>given time</u>.

2) You can compare how much energy is transferred by appliances with <u>different power ratings</u>.

> **EXAMPLE:** A 1.5 kW heater is left on for 1.5 hours.
> A 4 kW heater is also left on for 1.5 hours.
> Compare the energy transferred by both heaters.
>
> **ANSWER:** <u>Energy transferred (kWh) = power rating (kW) × time (h)</u>.
> Energy transferred by the 1.5 kW heater = 1.5 × 1.5 = 2.25 kWh.
> Energy transferred by the 4 kW heater = 4 × 1.5 = 6 kWh.
>
> Difference in the energy transferred by the two heaters = 6 − 2.25 = 3.75 kWh.
>
> So the 4 kW heater transfers <u>3.75 kWh more energy</u> than the 1.5 kW heater in 1.5 hours.

Food Labels Tell You How Much Energy is in Food

1) All the <u>food</u> we eat contains <u>energy</u>.

2) The energy in food is measured in <u>kilojoules (kJ)</u>.

3) You can <u>compare</u> the amount of <u>energy</u> in different foods by looking at their <u>labels</u>.

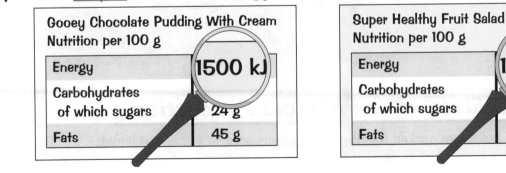

Gooey Chocolate Pudding With Cream Nutrition per 100 g	
Energy	1500 kJ
Carbohydrates of which sugars	24 g
Fats	45 g

Super Healthy Fruit Salad Nutrition per 100 g	
Energy	150 kJ
Carbohydrates of which sugars	9 g
Fats	0 g

What's my power rating you ask? About 8 out of 10...

Reading <u>labels</u> is not just good fun, but also useful — a label can tell you the power rating of an <u>appliance</u>, or how much <u>energy</u> is in your <u>food</u>. And as they say — energy makes the world go round. Ok, I might have made that up, but this stuff's <u>important</u> — make sure you <u>learn</u> it good and proper.

Physical Changes

Physical Changes Can Lead to a Change in State

1) A substance can either be a <u>solid</u>, a <u>liquid</u> or a <u>gas</u>. These are called <u>states of matter</u>.

2) A substance can <u>change</u> from one state of matter to another (see page **33**).
   For example, water can change from a liquid to a solid. This is called a <u>physical change</u>.

3) A physical change is <u>different</u> to a <u>chemical change</u>.
   In a <u>physical change</u> there is <u>no chemical reaction</u> and <u>no new substances</u> are made.
   For example, ice is still water — just in a different (physical) state.

4) You need to learn these <u>six different processes</u> that can bring about a physical change:

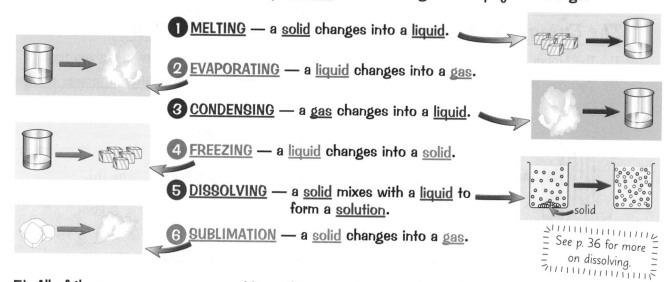

❶ <u>MELTING</u> — a <u>solid</u> changes into a <u>liquid</u>.

❷ <u>EVAPORATING</u> — a <u>liquid</u> changes into a <u>gas</u>.

❸ <u>CONDENSING</u> — a <u>gas</u> changes into a <u>liquid</u>.

❹ <u>FREEZING</u> — a <u>liquid</u> changes into a <u>solid</u>.

❺ <u>DISSOLVING</u> — a <u>solid</u> mixes with a <u>liquid</u> to form a <u>solution</u>.

❻ <u>SUBLIMATION</u> — a <u>solid</u> changes into a <u>gas</u>.

solid

See p. 36 for more on dissolving.

5) All of these processes are <u>reversible</u> — this means they can be '<u>undone</u>' by another process.
   For example, if you <u>melt</u> some <u>ice</u> you get <u>liquid water</u>. If you want to reverse this process
   you can <u>freeze</u> the liquid water to get ice again.

Physical Changes Don't Involve a Change in Mass

When a substance
changes from one state
of matter to another,
its <u>mass doesn't change</u>.

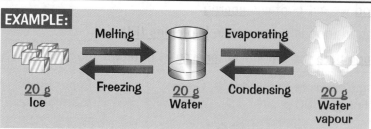

EXAMPLE:

Melting · Freezing · Evaporating · Condensing

20 g
Ice · 20 g
Water · 20 g
Water
vapour

Changes of State Alter Physical Properties

1) The particles in <u>solids</u> are <u>packed together tightly</u> compared to gases and liquids.
   This means solids have a <u>higher density</u> than gases and liquids.

2) When you <u>heat</u> a substance, it <u>changes</u> from a solid, to a liquid, to a gas.
   The particles move <u>further apart</u> and the substance becomes <u>less dense</u>.

3) <u>Ice</u> is different though. When you heat ice (so that it <u>melts</u>) the particles actually
   come <u>closer together</u>. So liquid water has a <u>higher density</u> than solid ice.

There's more about particles on p. 32.

All this studying's getting me in a right change of state...

Remember, physical changes <u>don't</u> create <u>new substances</u> or <u>change the mass</u> of a substance.

Movement of Particles

Particles can <u>move around</u> by <u>bashing</u> into each other and <u>bouncing off</u> in a new direction. Sounds like fun.

Brownian Motion is the Random Movement of Particles

1) <u>Brownian motion</u> is the <u>random movement</u> of any particle <u>suspended</u> (floating) within a <u>liquid</u> or <u>gas</u>.

2) It's a result of <u>collisions</u> between <u>particles</u>.

3) Remember — <u>atoms</u> and <u>molecules</u> are both types of <u>particle</u>. (See pages 34 and 35).

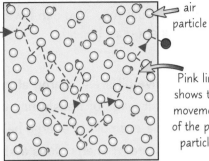

air particle

Pink line shows the movement of the pink particle.

Diffusion is Caused by the Random Motion of Particles

1) The particles in a liquid or gas move around at <u>random</u>.

2) Particles eventually bump their way from an area of <u>high concentration</u> (a place where there are lots of them) to an area of <u>low concentration</u> (a place where there aren't as many of them).

3) They constantly <u>bump</u> into each other, until they're <u>evenly spread</u> out across the substance.

High concentration of spectacle-particles   Low concentration of spectacle-particles

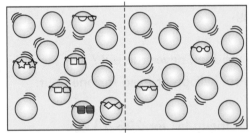

The particles move about randomly until...

...there's an even concentration of spectacle-particles on each side.

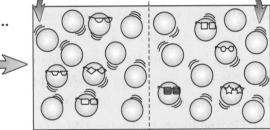

Movement of Particles Increases With Temperature

1) An <u>increase</u> in <u>temperature</u> causes particles to move around <u>more</u>.

2) So the <u>spaces between the particles</u> get <u>bigger</u>. This causes the substance to <u>expand</u>.

3) This explains how a <u>thermometer</u> works. Look:

EXAMPLE:

When a thermometer is heated the volume of liquid in the bulb <u>expands</u>. The particles move apart with their <u>extra energy</u>.

So the liquid moves <u>up</u> the thin tube of the <u>thermometer</u>.

I like to move it move it — you like to learn it learn it...

Brownian motion sounds tricky, but it's really not that bad when you think about it. Basically it's just particles <u>colliding</u> with each other all the time and <u>bouncing</u> off all over the place. You need to <u>learn</u> this, so cover up the page, scribble down what you remember and repeat until you know it all.

Section Summary

Ah, the section summary — on the home stretch at last.
In this section you've looked at the wonderful world of energy and matter.
All that's left for you now is to work through the exciting questions below.
You won't regret taking the time to work through these questions if you want to be super-amazing at science, trust me. You can have the odd cheeky peek back at the right page if you're stuck —
I won't tell anyone, honest.

1)   Name five different energy stores.

2)   Give four ways of transferring energy.

3)   Give one example of energy being transferred to a kinetic energy store.

4)   What store of energy does a stretched slingshot transfer energy from when it's released?

5)   Describe one example of a force doing work and transferring energy.

6)* Two identical cranes each transfer 20 kJ of energy to move a weight. One crane applies a big force, the other applies a small force. Which crane can lift the weight the furthest?

7)   What does thermal equilibrium mean?

8)   Name two ways in which energy can be transferred by heating between two objects.
     Briefly describe how energy is transferred in each of these ways.

9)   How does adding an insulator to an object affect the rate of energy transfer?

10) True or false? Energy is sometimes destroyed.

11) Why are most energy transfers **NOT** perfect?

12) Give two energy resources created using the Sun's energy.

13) Give an example of a non-renewable energy resource.

14) Give two examples of renewable energy resources. Why will they never run out?

15)* Calculate the energy transferred by a 1.5 kW remote-control helicopter used for one hour.

16) What unit is household electricity measured in?

17)* Electricity costs 15p per kWh. Calculate the cost of an electricity bill for 298.2 kWh.

18) What does the power rating of an appliance tell you?

19)* Which will transfer more energy — a 200 W device left on for 1 hour, or a 300 W device left on for 1 hour?

20) What unit is the energy in food usually measured in?

21) Where could you find out the energy contained in a packet of chocolate-covered sugar cubes?

22) Why is a physical change different to a chemical change?

23) Melting is a process that brings about a physical change.
     Name five other processes that bring about a change of state.

24) True or false? Melting is not a reversible process.

25)* 50 g of iron is melted. How much liquid iron would be produced?

26) Give one difference between the physical properties of a gas and a solid of the same substance.

27) What's meant by Brownian motion?

28) Particles in gases and liquids moves from areas of high concentration to low concentration.
     What name is given to this process?

29) Explain why gases expand when they're heated.

*Answers on page 108.

Speed

Neeeeoww... Yes, it's a page on speed. Make sure you can do these <u>calculations</u> — don't <u>zoom off</u>.

Speed is How Fast You're Going

1) <u>Speed</u> is a <u>measure</u> of how <u>far</u> you travel in a <u>set</u> amount of <u>time</u>.

2) You can use this formula to do <u>speed calculations</u>:

$$\text{Speed} = \frac{\text{Distance}}{\text{Time}}$$

This line means divided by (÷).

3) You can use the word <u>SIDOT</u> to help you remember the formula:

<u>SIDOT</u> — <u>S</u>peed <u>I</u>s <u>D</u>istance <u>O</u>ver <u>T</u>ime.

4) These are <u>three</u> common <u>units</u> for speed.

<u>metres</u> per <u>second</u> — m/s
<u>miles</u> per <u>hour</u> — mph or miles/h
<u>kilometres</u> per <u>hour</u> — km/h

Always use <u>UNITS</u>.

Work Out Speed Using Distance and Time

To work out <u>SPEED</u> you need to know the <u>distance travelled</u> and the <u>time taken</u>.

Example 1:

A hooligan sheep is skateboarding down a farmer's track.
The sheep takes exactly <u>5 seconds</u> to move between two fence posts.
The posts are <u>10 metres</u> apart. <u>What is the sheep's SPEED?</u>

10 m

Answer:

<u>STEP 1)</u>   <u>Write down what you know</u>:
            distance = 10 m   time = 5 s

<u>STEP 2)</u>   <u>Use the formula</u>:
            Speed = Distance ÷ Time   = 10 ÷ 5   = <u>2 m/s</u>

The answer is in metres per second (m/s) because the distance was given in metres and the time in seconds.

Distance-Time Graphs

A distance-time graph shows the <u>distance</u> travelled by an object over <u>time</u>.

1) The <u>slope</u> of the line (<u>gradient</u>) shows the <u>speed</u> at which the object is moving.

2) A <u>steeper</u> line means the object is moving <u>faster</u>.

3) <u>Flat</u> sections are where it's <u>stopped</u>.

4) <u>Downhill</u> sections mean it's moving <u>back</u> toward its <u>starting point</u>.

5) <u>Curves</u> mean the speed is <u>changing</u>.

6) A curve that gets <u>steeper</u> means the object is <u>speeding up</u> (<u>accelerating</u>).

7) A curve <u>levelling off</u> means the object is <u>slowing down</u> (<u>decelerating</u>).

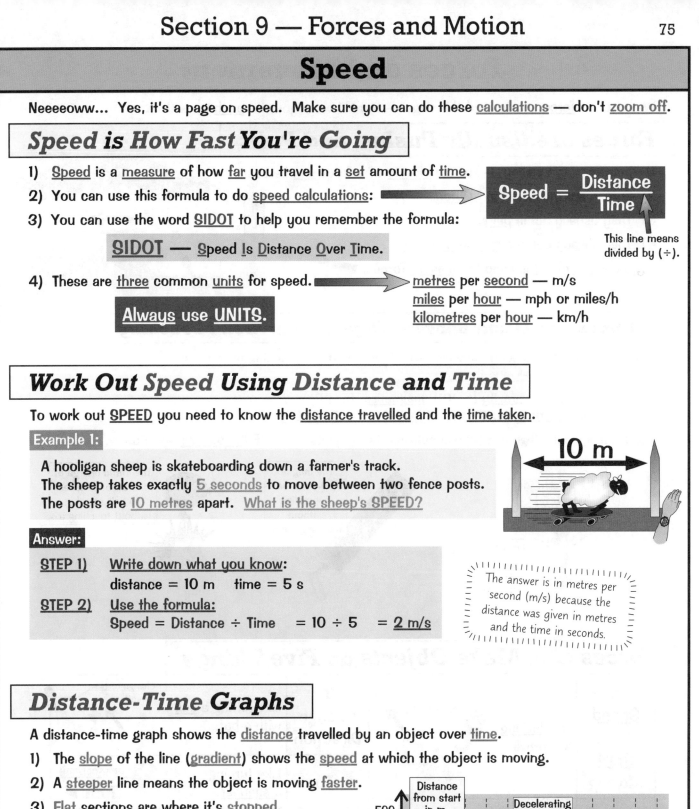

Speed is ace — well it takes some beating...

Speed is a pretty simple idea really. This page has all the <u>really basic</u> and important facts about <u>speed</u>. There's the formula for a start, and the units, and then a worked example. <u>Learn it all</u>. Now.

Forces and Movement

Well, I can't <u>force</u> you to read this page — but if I were you, I'd <u>push</u> on with it...

Forces are Usually Pushes and Pulls

1) Forces <u>can't</u> be seen — but the <u>effects</u> of a force <u>can</u> be seen.
2) Forces are measured in <u>Newtons</u> — <u>N</u>.
3) They usually act in <u>pairs</u>.
4) They <u>always</u> act in a <u>certain direction</u>.
5) A <u>newton meter</u> is used to <u>measure</u> forces.

Forces Can Occur Between Objects That Aren't Touching

1) Forces usually occur between two objects that are <u>touching each other</u>. But this <u>isn't always the case</u>.
2) Forces due to <u>gravity</u> (p. 101), <u>magnetism</u> (p. 98) and <u>static electricity</u> (p. 97) are all <u>NON-CONTACT FORCES</u>.
3) This means they can happen between two objects that <u>AREN'T</u> touching each other.

Gravity makes the elephant fall without touching it.

The magnet attracts the paperclips without touching them.

The static charge of the balloon makes the hair stand up without touching it.

Forces Can Make Objects do Five Things

1. <u>Speed</u> Up or <u>Start</u> Moving	Like <u>kicking</u> a football.	3. Change Direction	Like hitting a <u>ball</u> with a <u>bat</u>.
		4. Turn	Like <u>turning</u> <u>a spanner</u>.
2. <u>Slow</u> Down or <u>Stop</u> Moving	Like <u>air resistance</u> (see next page).	5. Change Shape	Like <u>stretching</u> and <u>compressing</u> (see p. 80), <u>bending</u> and <u>twisting</u>.

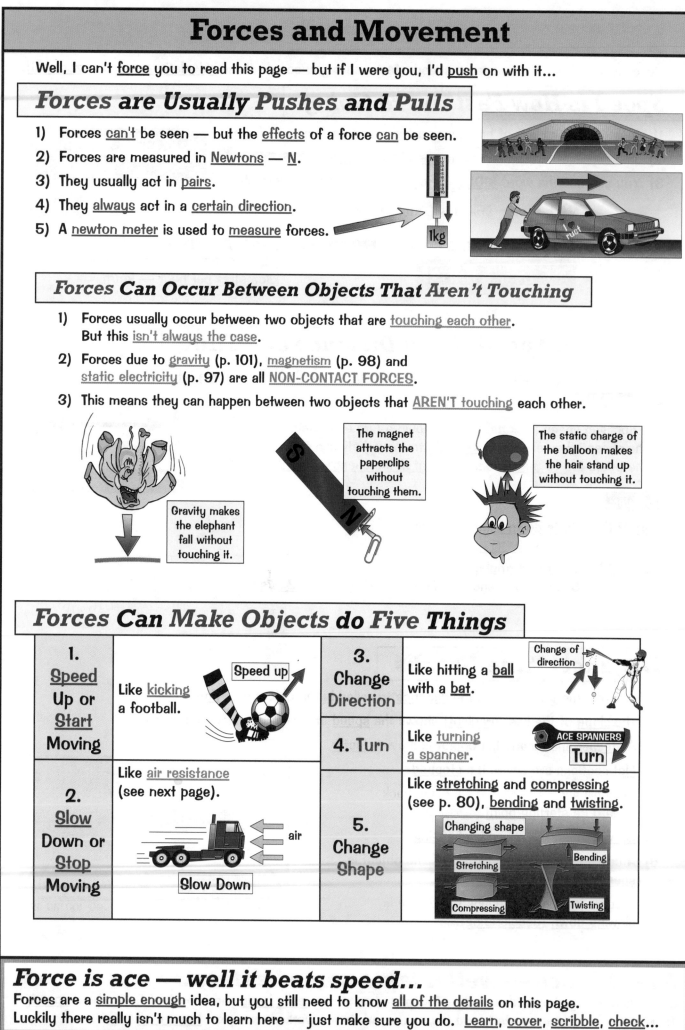

Force is ace — well it beats speed...

Forces are a <u>simple enough</u> idea, but you still need to know <u>all of the details</u> on this page. Luckily there really isn't much to learn here — just make sure you do. <u>Learn</u>, <u>cover</u>, <u>scribble</u>, <u>check</u>...

Friction and Resistance

Friction and resistance are good at <u>slowing things down</u>. Don't let them turn studying into a drag.

Friction Tries to Stop Objects Sliding Past Each Other

1) Friction is a <u>force</u> that always acts in the <u>opposite</u> direction to movement.

2) To <u>start</u> something moving, a push or pull force must be <u>bigger</u> than <u>resisting forces</u> like friction.

3) So to <u>push an object</u> out of the way, you need to <u>overcome</u> friction.

4) You get friction when:
   - two surfaces <u>rub together</u>.
   - an object <u>passes through air</u> or <u>water</u>.

Air and Water Resistance Slow Down Moving Objects

1) Air and water resistance are <u>frictional forces</u>.

2) They <u>push against</u> objects which are moving through air or water and <u>slow</u> them down.

How Air Resistance Affects Sheep Jumping Out of Planes

(It happens all the time round here.)

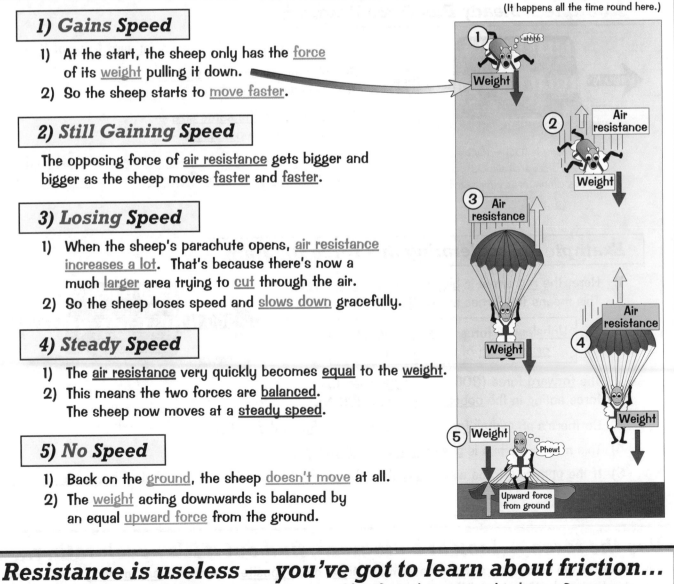

1) Gains Speed

1) At the start, the sheep only has the <u>force</u> of its <u>weight</u> pulling it down.

2) So the sheep starts to <u>move faster</u>.

2) Still Gaining Speed

The opposing force of <u>air resistance</u> gets bigger and bigger as the sheep moves <u>faster</u> and <u>faster</u>.

3) Losing Speed

1) When the sheep's parachute opens, <u>air resistance</u> <u>increases a lot</u>. That's because there's now a much <u>larger</u> area trying to <u>cut</u> through the air.

2) So the sheep loses speed and <u>slows down</u> gracefully.

4) Steady Speed

1) The <u>air resistance</u> very quickly becomes <u>equal</u> to the <u>weight</u>.

2) This means the two forces are <u>balanced</u>. The sheep now moves at a <u>steady speed</u>.

5) No Speed

1) Back on the <u>ground</u>, the sheep <u>doesn't move</u> at all.

2) The <u>weight</u> acting downwards is balanced by an equal <u>upward force</u> from the ground.

Resistance is useless — you've got to learn about friction...

... as well as air and water resistance. Here's a page of <u>key facts</u> that you need to <u>learn</u>. Go go go.

Force Diagrams

Force diagrams. They're <u>diagrams</u> that show <u>forces</u>. Bet you weren't expecting that...

Use Force Diagrams to Show the Forces Acting on Objects

Example 1: Stationary Teapot Force Diagram

A <u>stationary</u> object <u>doesn't move</u>. This <u>teapot</u> is <u>stationary</u>.

1) The teapot's <u>weight</u> is a <u>force</u> acting <u>downwards</u> on the table. It's the <u>red</u> arrow.

2) A <u>reaction force</u> from the table's surface <u>pushes up</u> on the teapot. This is the <u>blue</u> arrow.

3) The <u>reaction force</u> and <u>weight</u> are <u>EQUAL</u> and <u>OPPOSITE</u>. The <u>arrows</u> are the <u>same size</u> and point in <u>opposite directions</u>.

4) This means the <u>forces</u> on the teapot are <u>BALANCED</u>. So it stays <u>stationary</u> (not moving).

5) That's because: | <u>Balanced forces</u> produce <u>no change</u> in movement. |

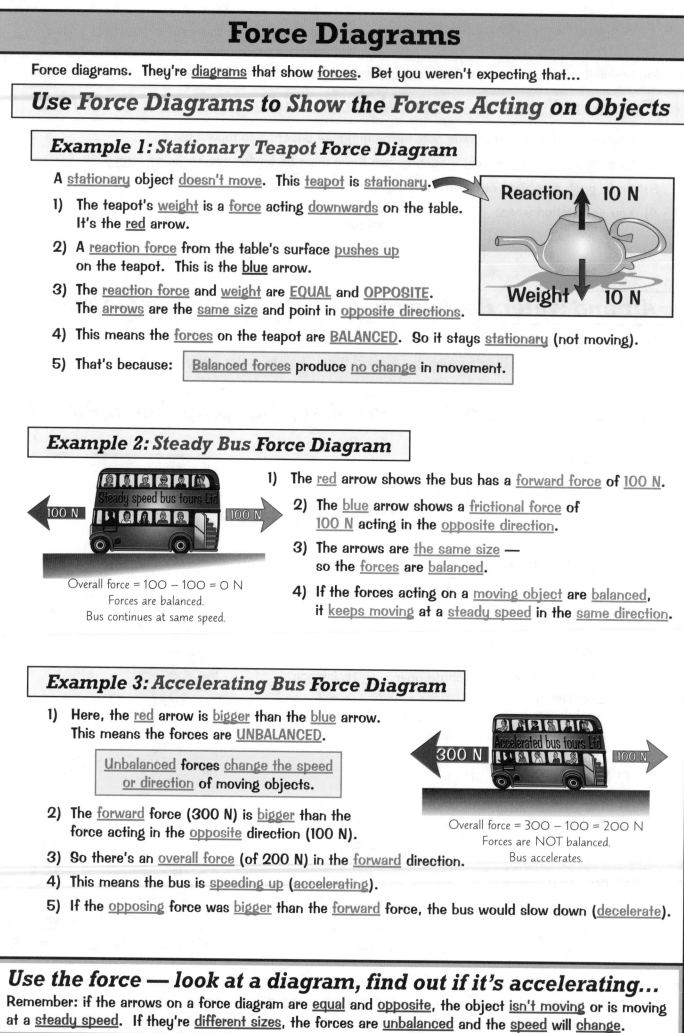

Reaction ↑ 10 N

Weight ↓ 10 N

Example 2: Steady Bus Force Diagram

100 N ← Steady speed bus tours Ltd → 100 N

Overall force = 100 − 100 = 0 N
Forces are balanced.
Bus continues at same speed.

1) The <u>red</u> arrow shows the bus has a <u>forward force</u> of <u>100 N</u>.

2) The <u>blue</u> arrow shows a <u>frictional force</u> of <u>100 N</u> acting in the <u>opposite direction</u>.

3) The arrows are <u>the same size</u> — so the <u>forces</u> are <u>balanced</u>.

4) If the forces acting on a <u>moving object</u> are <u>balanced</u>, it <u>keeps moving</u> at a <u>steady speed</u> in the <u>same direction</u>.

Example 3: Accelerating Bus Force Diagram

1) Here, the <u>red</u> arrow is <u>bigger</u> than the <u>blue</u> arrow. This means the forces are <u>UNBALANCED</u>.

| <u>Unbalanced</u> forces <u>change the speed or direction</u> of moving objects. |

300 N ← Accelerated bus tours Ltd → 100 N

Overall force = 300 − 100 = 200 N
Forces are NOT balanced.
Bus accelerates.

2) The <u>forward</u> force (300 N) is <u>bigger</u> than the force acting in the <u>opposite</u> direction (100 N).

3) So there's an <u>overall force</u> (of 200 N) in the <u>forward</u> direction.

4) This means the bus is <u>speeding up</u> (<u>accelerating</u>).

5) If the <u>opposing</u> force was <u>bigger</u> than the <u>forward</u> force, the bus would slow down (<u>decelerate</u>).

Use the force — look at a diagram, find out if it's accelerating...

Remember: if the arrows on a force diagram are <u>equal</u> and <u>opposite</u>, the object <u>isn't moving</u> or is moving at a <u>steady speed</u>. If they're <u>different sizes</u>, the forces are <u>unbalanced</u> and the <u>speed</u> will <u>change</u>.

Moments

Don't wait a lifetime to learn moments like this — <u>memorise</u> what's on this page <u>now</u>.

Forces Cause Objects to Turn Around Pivots

A <u>pivot</u> is the point around which rotation happens — like the middle of a <u>seesaw</u>.

A Moment is the Turning Effect of a Force

1) When a <u>force acts</u> on something which has a <u>pivot</u>, it creates a <u>moment</u> (turning effect).

2) Learn this important equation:

Moment = force x distance

in newton metres, Nm   in newtons, N   in metres, m

Balancing Moments

Balanced moments mean that...

anticlockwise moments = clockwise moments

If the moments are <u>not</u> balanced, the ruler will turn in the direction of the bigger moment.

Clockwise

<u>ANTICLOCKWISE</u>   force x distance   =   force x distance   <u>CLOCKWISE</u>

100 N x 0.5 m   =   100 N x 0.5 m

<u>50 Nm</u>   =   <u>50 Nm</u>   — BALANCED

50 cm = 0.5 m

Is it Balanced?

You can calculate moments to work out what will happen to these beams...

1) <u>ANTICLOCKWISE</u>:   <u>CLOCKWISE</u>:

5 × 3 = 15 Nm   **5 × 3 = 15 Nm**

The anticlockwise moment <u>EQUALS</u> the clockwise moment. So the beam is <u>BALANCED</u>.

2) <u>ANTICLOCKWISE</u>:   <u>CLOCKWISE</u>:

6 × 1 = 6 Nm   **2 × 2 = 4 Nm**

The <u>anticlockwise</u> moment is <u>bigger</u> than the clockwise moment. The beam is <u>NOT BALANCED</u>.

It will <u>TURN</u> in the <u>ANTICLOCKWISE DIRECTION</u>.

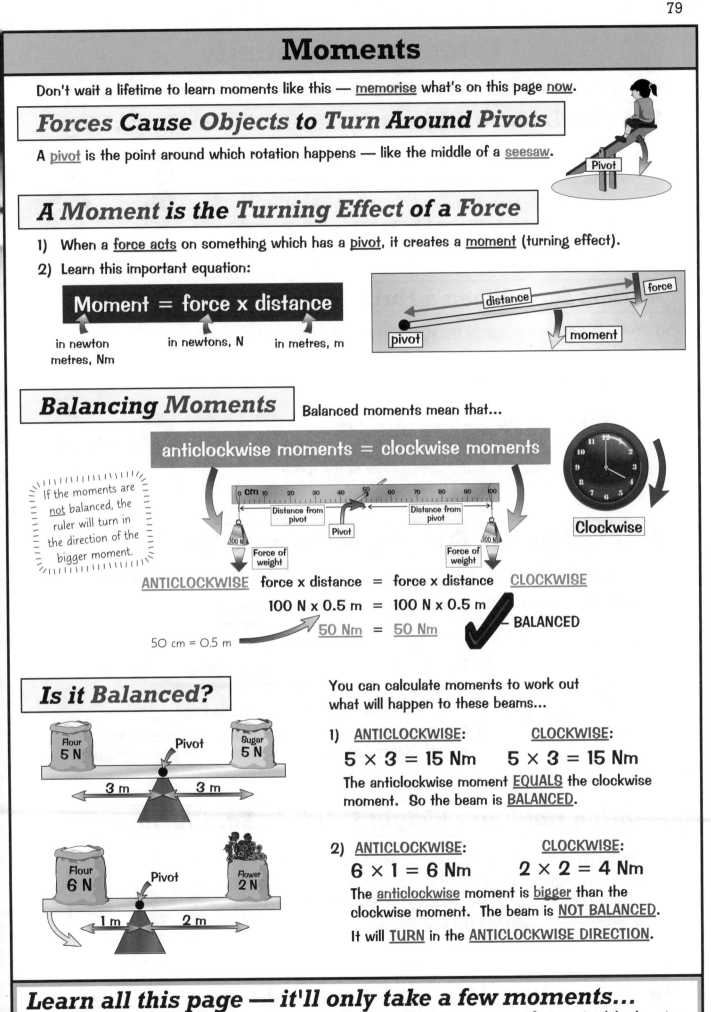

Learn all this page — it'll only take a few moments...

"Moment" is a bit of a <u>weird name</u> really. Seems like it's designed to cause <u>confusion</u>. Luckily, learning this page good and proper will banish any confusion to the <u>dustbin of doom</u>. So read, cover, scribble...

Forces and Elasticity

It's not just about turning, pushing and pulling. Forces are also able to <u>stretch</u> or <u>squash</u> things.

You Can Deform Objects by Stretching or Squashing

1) You can use forces to <u>stretch</u> or <u>compress</u> (squash) objects.

2) The force you apply causes the object to <u>deform</u> (change its shape).

3) <u>Springs</u> are <u>special</u> because they usually <u>spring back</u> into their <u>original shape</u> after the force has been <u>removed</u>. They are <u>elastic</u>.

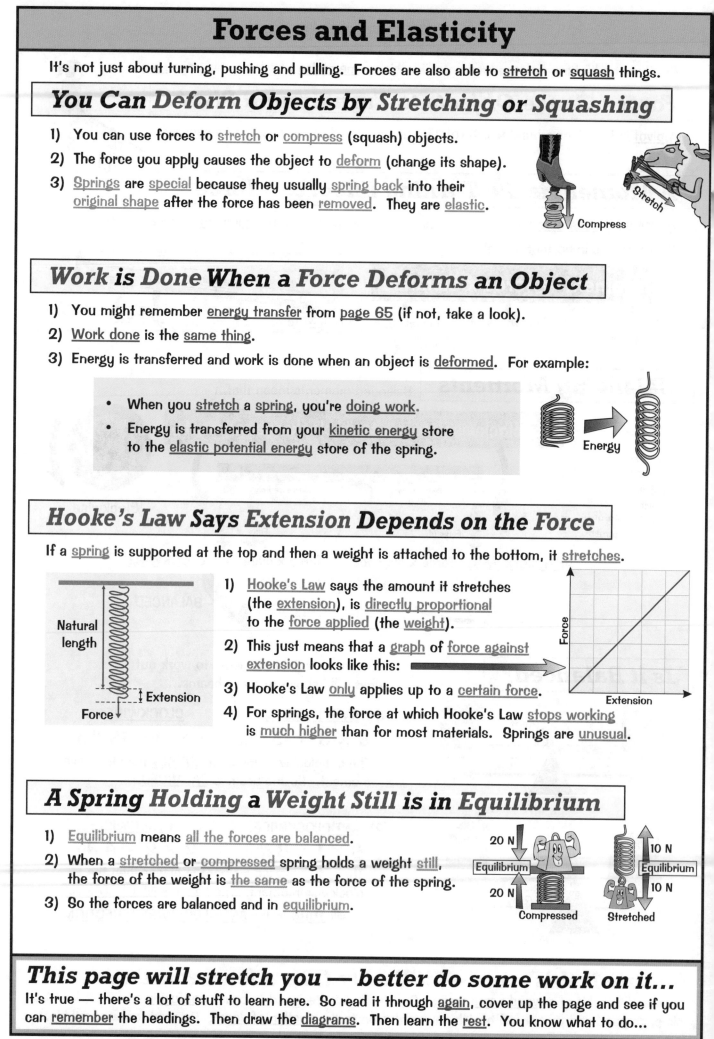

Compress

Stretch

Work is Done When a Force Deforms an Object

1) You might remember <u>energy transfer</u> from <u>page 65</u> (if not, take a look).

2) <u>Work done</u> is the <u>same thing</u>.

3) Energy is transferred and work is done when an object is <u>deformed</u>. For example:

- When you <u>stretch</u> a <u>spring</u>, you're <u>doing work</u>.
- Energy is transferred from your <u>kinetic energy</u> store to the <u>elastic potential energy</u> store of the spring.

Energy

Hooke's Law Says Extension Depends on the Force

If a <u>spring</u> is supported at the top and then a weight is attached to the bottom, it <u>stretches</u>.

Natural length

Extension

Force

1) <u>Hooke's Law</u> says the amount it stretches (the <u>extension</u>), is <u>directly proportional</u> to the <u>force applied</u> (the <u>weight</u>).

2) This just means that a <u>graph</u> of <u>force against extension</u> looks like this:

3) Hooke's Law <u>only</u> applies up to a <u>certain force</u>.

4) For springs, the force at which Hooke's Law <u>stops working</u> is <u>much higher</u> than for most materials. Springs are <u>unusual</u>.

Force

Extension

A Spring Holding a Weight Still is in Equilibrium

1) <u>Equilibrium</u> means <u>all the forces are balanced</u>.

2) When a <u>stretched</u> or <u>compressed</u> spring holds a weight <u>still</u>, the force of the weight is <u>the same</u> as the force of the spring.

3) So the forces are balanced and in <u>equilibrium</u>.

20 N
Equilibrium
20 N
Compressed

10 N
Equilibrium
10 N
Stretched

This page will stretch you — better do some work on it...

It's true — there's a lot of stuff to learn here. So read it through <u>again</u>, cover up the page and see if you can <u>remember</u> the headings. Then draw the <u>diagrams</u>. Then learn the <u>rest</u>. You know what to do...

Pressure

Don't let pressure <u>get you down</u> — here's a lovely page that explains it all.

Pressure *is How Much Force is Put on a Certain Area*

1) The formula shows how <u>pressure</u>, <u>force</u> and <u>area</u> are linked:

2) A given force acting over a <u>BIG area</u> means a <u>small pressure</u>.

3) A given force acting over a <u>small area</u> means a <u>BIG pressure</u>.

$$\text{Pressure} = \frac{\text{Force}}{\text{Area}}$$

This line means divided by (÷).

Pressure *is measured in* N/m² *or Pascals* (Pa)

If a force of <u>1 Newton</u> is spread over an area of <u>1 m²</u> (like this) then it applies a pressure of <u>1 Pascal</u>.

$$1 \text{ Newton/metre}^2 = 1 \text{ Pascal}$$
$$1 \text{ N/m}^2 = 1 \text{ Pa}$$

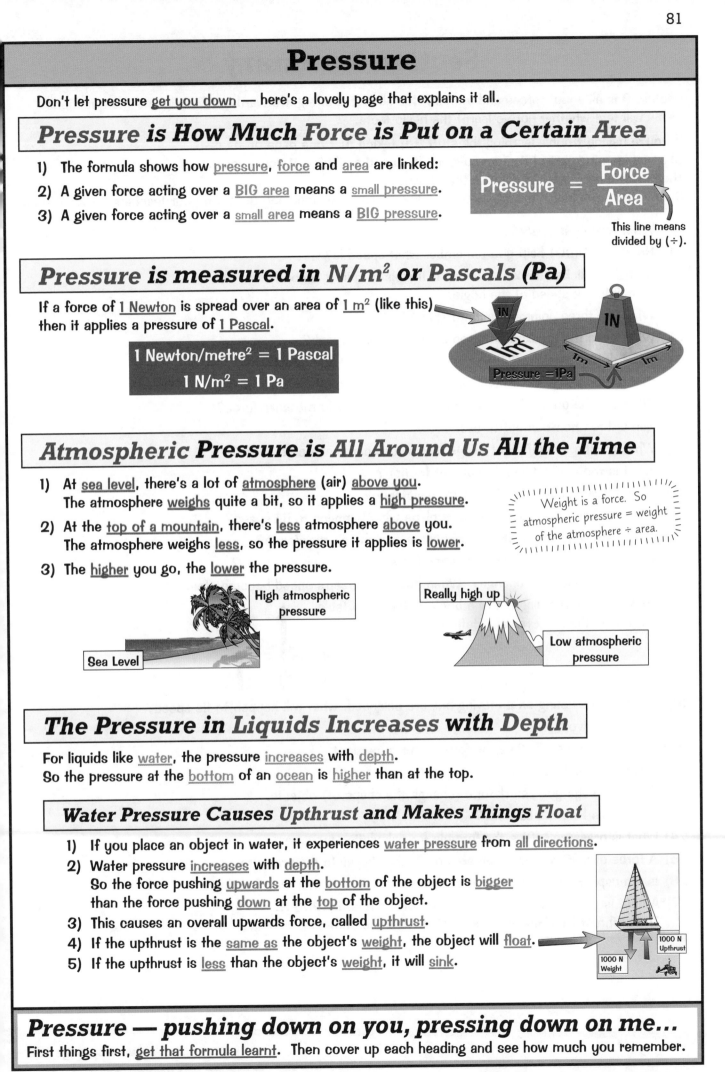

1N

1N

1m²

Pressure = 1Pa

1m   1m

Atmospheric *Pressure is All Around Us All the Time*

1) At <u>sea level</u>, there's a lot of <u>atmosphere</u> (air) <u>above you</u>. The atmosphere <u>weighs</u> quite a bit, so it applies a <u>high pressure</u>.

2) At the <u>top of a mountain</u>, there's <u>less</u> atmosphere <u>above</u> you. The atmosphere weighs <u>less</u>, so the pressure it applies is <u>lower</u>.

3) The <u>higher</u> you go, the <u>lower</u> the pressure.

> Weight is a force. So atmospheric pressure = weight of the atmosphere ÷ area.

High atmospheric pressure

Sea Level

Really high up

Low atmospheric pressure

The Pressure in Liquids Increases *with Depth*

For liquids like <u>water</u>, the pressure <u>increases</u> with <u>depth</u>.
So the pressure at the <u>bottom</u> of an <u>ocean</u> is <u>higher</u> than at the top.

Water Pressure Causes Upthrust *and Makes Things Float*

1) If you place an object in water, it experiences <u>water pressure</u> from <u>all directions</u>.

2) Water pressure <u>increases</u> with <u>depth</u>. So the force pushing <u>upwards</u> at the <u>bottom</u> of the object is <u>bigger</u> than the force pushing <u>down</u> at the <u>top</u> of the object.

3) This causes an overall upwards force, called <u>upthrust</u>.

4) If the upthrust is the <u>same as</u> the object's <u>weight</u>, the object will <u>float</u>.

5) If the upthrust is <u>less</u> than the object's <u>weight</u>, it will <u>sink</u>.

1000 N Upthrust

1000 N Weight

Pressure *— pushing down on you, pressing down on me...*

First things first, <u>get that formula learnt</u>. Then cover up each heading and see how much you remember.

Section Summary

Section 9 is all about forces and motion. It's all pretty straightforward stuff really and the questions below will test whether you've learnt the basic facts.

If you're having trouble learning the stuff, try taking just one page on its own.
Start by learning part of it, then covering it up and scribbling it down again.
Then learn a bit more and scribble that down.
Soon you'll have learnt the whole section and be ready to face any question your teachers throw at you.

1) What exactly is speed?

2) How does SIDOT help you remember what speed is?

3)* A bogie is flicked across a room. It travels 5 m in 2 seconds.
   Calculate the speed of the bogie.

4) What are three common units of speed?

5) What does the gradient show on a distance-time graph?

6) What does a straight, flat line mean on a distance-time graph?

7) Can forces be seen? How do we know they're there?

8) What are the units of force? What would you use to measure force?

9) Name three non-contact forces.

10) What are the five different things that forces can make objects do?

11) What is friction? When do you get friction?

12) What is air resistance? And water resistance?

13) When a sheep first jumps out of a plane what happens to its speed?

14) As the sheep moves faster, what happens to the air resistance?

15) What happens to air resistance when the sheep's parachute opens?

16) Does the speed then change? When does the sheep's speed become steady?

17) Look at the force diagram of a book resting on a table.
    Why does the book remain stationary?

18) If the forces acting on a moving bus are balanced, what will happen to its speed?

19) What is a pivot?

20) What is a moment? Give the formula for a moment.

21) What does "balanced moments" mean?

22) Give two ways you can deform (change the shape of) objects.

23) What does Hooke's Law say?

24) What is pressure? Give the formula for calculating pressure.

25)*A force of 200 N acts on an area of 2 m². Calculate the pressure.

26) Is atmospheric pressure higher at the seaside or up a mountain? Why?

27)*A boat is put into the sea. The weight of the boat is 1000 N.
    The upthrust from the water is 800 N. Will the boat sink or float?

*Answers on page 108

Water Waves

Take a <u>deep breath</u> and <u>dive into</u> the wonderful world of water waves...

Water Waves are Transverse

1) <u>Waves</u> travelling across the <u>ocean</u> are <u>transverse waves</u>.

2) A transverse wave has <u>undulations</u> (<u>up</u> and <u>down</u> movements).
   These movements are at <u>right angles</u> to the <u>direction</u> the wave is travelling in.

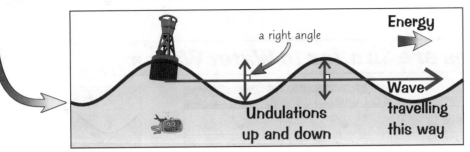

a right angle

Energy

Undulations up and down

Wave travelling this way

3) Waves <u>transfer energy</u> from one place to another.
   Energy is transferred in the direction the wave is travelling in.

Waves Can be Reflected

1) If a water wave hits a surface, it will be <u>reflected</u>.

2) This causes the <u>direction</u> of the wave to change.

3) <u>All waves</u> can be reflected.
   There's more on reflection on pages 85 and 90.

Incoming wave

Reflected wave

Sea Wall

Sea Wall

Transverse Waves Have Crests and Troughs

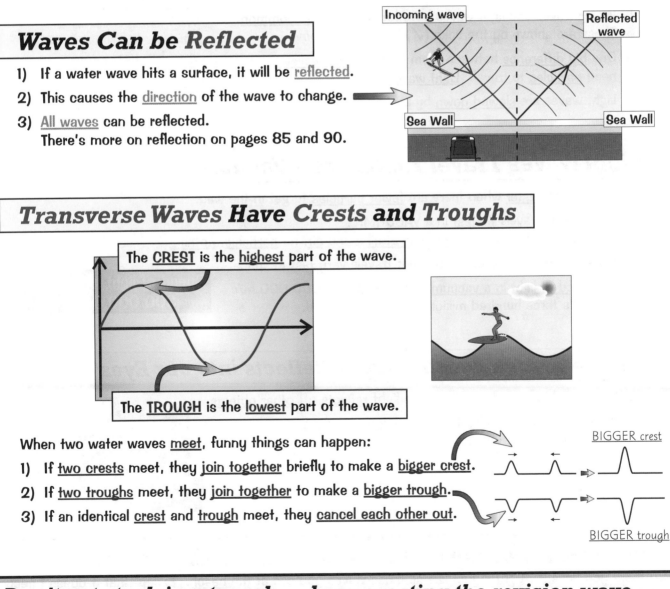

The <u>CREST</u> is the <u>highest</u> part of the wave.

The <u>TROUGH</u> is the <u>lowest</u> part of the wave.

When two water waves <u>meet</u>, funny things can happen:

1) If <u>two crests</u> meet, they <u>join together</u> briefly to make a <u>bigger crest</u>.

2) If <u>two troughs</u> meet, they <u>join together</u> to make a <u>bigger trough</u>.

3) If an identical <u>crest</u> and <u>trough</u> meet, they <u>cancel each other out</u>.

BIGGER crest

BIGGER trough

Don't get stuck in a trough — keep cresting the revision wave...

<u>Write</u> down all the <u>technical words</u> on this page and what they <u>mean</u>. Then draw and label the <u>diagrams</u>.

Light Waves

You wouldn't know from looking, but light is actually a wave. Here's a page all about it...

Light is a Wave of Energy

1) Light is produced by luminous objects.
   These include the Sun, candles, light bulbs and flames.

2) Light is a wave. It always travels in a straight line.

Light Waves are Similar to Water Waves

	Water Wave	Light Wave
Is a transverse wave	✓	✓
Transfers energy	✓	✓
Can be reflected	✓	✓
Needs particles to travel	✓	✗

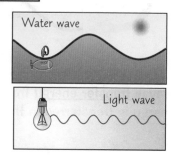

Water waves and light waves have the same shape — they're both transverse waves.

1) Water waves and light waves have lots of things in common. These are shown by the ticks (✓) in the table above.

2) One big difference between them is that water waves need particles to travel. Light waves DON'T.

3) Light waves are slowed down by particles.

Light Waves Travel Fastest in a Vacuum

1) Light travels faster when there are fewer particles to get in the way.

2) Light always travels fastest in a VACUUM.
   A vacuum is where there is nothing at all — no air, no particles, nothing.

3) Space is mostly a vacuum. So light travels very fast in space.

4) The speed of light in a vacuum is ALWAYS 300 000 000 m/s
   — that's three hundred million metres per second.

> Speed of light in a vacuum
> 300 000 000 m/s

We See Things Because Light Reflects into our Eyes

1) When an object produces light, the light reflects off other objects.

2) Some of the reflected light then goes into our eyes.
   This is how we see (see page 87).

CGP Top Value Book

Person | Reflects off book into eager eyes | Light ray | Light source

Confused? Let me shine some light on the problem...

So there you have it. Light is just like all those waves you see at the beach. Except that it doesn't need a load of water to get from A to B. Anything like water puts a load of particles in the way and slows the light waves down. Nope, light only hits top speed when it's in a vacuum with nothing in the way at all.

Reflection

Reflection is all about what happens to light rays when they hit something. It's quite a sight...

Mirrors Have Shiny Surfaces Which Reflect Light

1) A light wave is also known as a light ray.

2) Light rays reflect off mirrors and most other things.

3) Mirrors have a very smooth shiny surface.

4) The shiny surface reflects all the light off at the same angle, giving a clear reflection. This is called SPECULAR REFLECTION.

5) Rough surfaces look dull because the light is reflected back (scattered) in lots of different directions. This is called DIFFUSE SCATTERING.

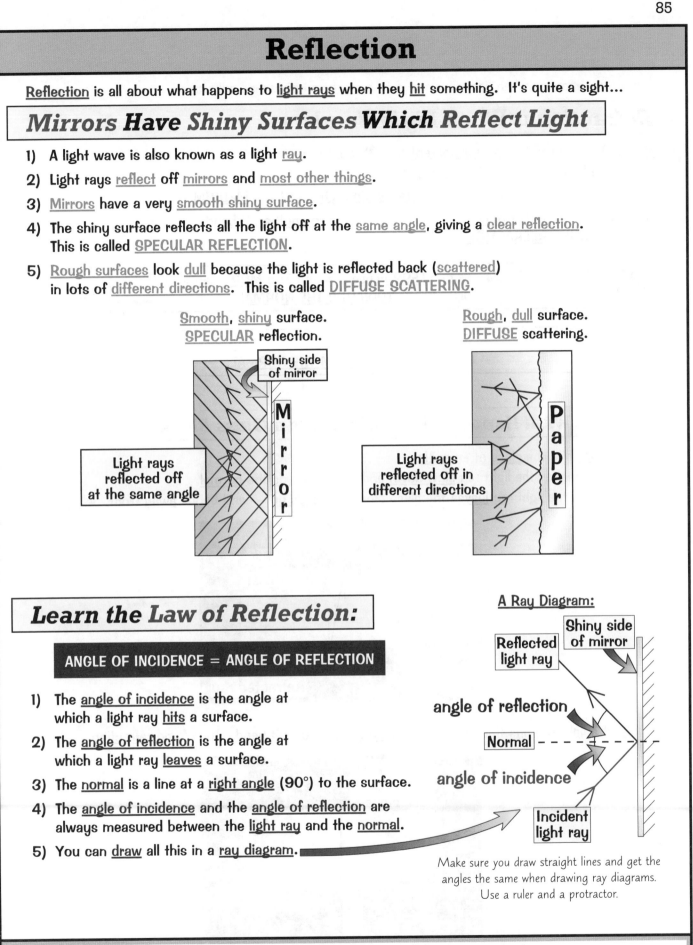

Smooth, shiny surface.
SPECULAR reflection.

Shiny side of mirror

Mirror

Light rays reflected off at the same angle

Rough, dull surface.
DIFFUSE scattering.

Paper

Light rays reflected off in different directions

Learn the Law of Reflection:

ANGLE OF INCIDENCE = ANGLE OF REFLECTION

1) The angle of incidence is the angle at which a light ray hits a surface.

2) The angle of reflection is the angle at which a light ray leaves a surface.

3) The normal is a line at a right angle (90°) to the surface.

4) The angle of incidence and the angle of reflection are always measured between the light ray and the normal.

5) You can draw all this in a ray diagram.

A Ray Diagram:

Shiny side of mirror

Reflected light ray

angle of reflection

Normal - - - - -

angle of incidence

Incident light ray

Make sure you draw straight lines and get the angles the same when drawing ray diagrams. Use a ruler and a protractor.

Normally you might grumble that this is dull...

...But if you take a moment to reflect on this page, it'll all become clear. First off, get those terms 'specular reflection' and 'diffuse scattering' learnt by covering the page and writing out their definitions. The Law of Reflection is really important, so make sure you've got it memorised. Then draw your own ray diagram — make sure the angle of reflection is the same as the angle of incidence. Hoo-ray.

Refraction

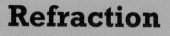

Refraction happens when light rays move to a more or less dense substance.

Refraction is When Light Bends

1) Light will travel through transparent (see-through) material.
   It won't go through anything opaque (not see-through).

2) Any substance that light or sound travels through is called a MEDIUM.

3) When light travels from one transparent medium to another, it bends.
   This is called REFRACTION.

LEARN THESE:

| When light goes from a LESS dense medium to a MORE dense medium: light bends TOWARDS THE NORMAL. |

Example: air to glass.

| When light goes from a MORE dense medium to a LESS dense medium: light bends AWAY FROM THE NORMAL. |

Example: glass to air.

Here's What Happens When Light Hits a Glass Block

1) Light hits the glass at an angle.
2) This makes the light ray slow down and bend towards the normal.
3) When the light ray leaves the glass block it speeds up — and bends away from the normal.

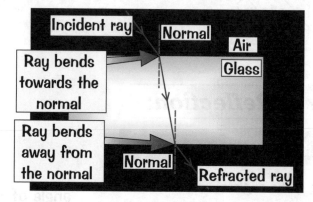

Glass is much more dense than air.

4) If a light ray hits the glass block straight on, it doesn't bend. There's no refraction.

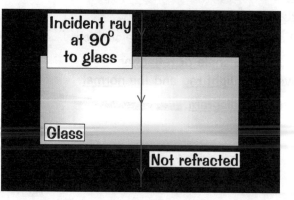

I used to be a medium — before I ate a lot of doughnuts...

Refraction is very different to reflection — the first thing you've got to do is spot that they're actually two different words. Ref-lec-tion and ref-rac-tion. Now all you need to do is learn what they both are.

Lenses and Cameras

An <u>important</u> page this one — it's all about how eyes and our cameras produce <u>images</u>.

The Pinhole Camera is a Simple Camera

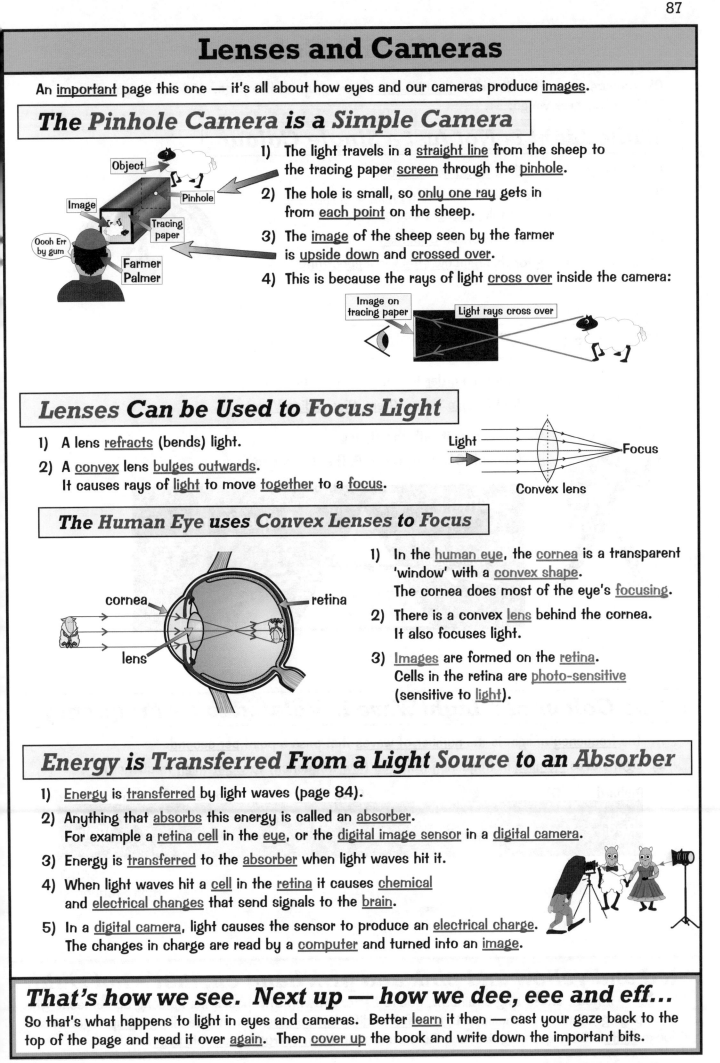

1) The light travels in a <u>straight line</u> from the sheep to the tracing paper <u>screen</u> through the <u>pinhole</u>.

2) The hole is small, so <u>only one ray</u> gets in from <u>each point</u> on the sheep.

3) The <u>image</u> of the sheep seen by the farmer is <u>upside down</u> and <u>crossed over</u>.

4) This is because the rays of light <u>cross over</u> inside the camera:

Lenses Can be Used to Focus Light

1) A lens <u>refracts</u> (bends) light.

2) A <u>convex</u> lens <u>bulges outwards</u>. It causes rays of <u>light</u> to move <u>together</u> to a <u>focus</u>.

The Human Eye uses Convex Lenses to Focus

1) In the <u>human eye</u>, the <u>cornea</u> is a transparent 'window' with a <u>convex shape</u>. The cornea does most of the eye's <u>focusing</u>.

2) There is a convex <u>lens</u> behind the cornea. It also focuses light.

3) <u>Images</u> are formed on the <u>retina</u>. Cells in the retina are <u>photo-sensitive</u> (sensitive to <u>light</u>).

Energy is Transferred From a Light Source to an Absorber

1) <u>Energy</u> is <u>transferred</u> by light waves (page 84).

2) Anything that <u>absorbs</u> this energy is called an <u>absorber</u>. For example a <u>retina cell</u> in the <u>eye</u>, or the <u>digital image sensor</u> in a <u>digital camera</u>.

3) Energy is <u>transferred</u> to the <u>absorber</u> when light waves hit it.

4) When light waves hit a <u>cell</u> in the <u>retina</u> it causes <u>chemical</u> and <u>electrical changes</u> that send signals to the <u>brain</u>.

5) In a <u>digital camera</u>, light causes the sensor to produce an <u>electrical charge</u>. The changes in charge are read by a <u>computer</u> and turned into an <u>image</u>.

That's how we see. Next up — how we dee, eee and eff...

So that's what happens to light in eyes and cameras. Better <u>learn</u> it then — cast your gaze back to the top of the page and read it over <u>again</u>. Then <u>cover up</u> the book and write down the important bits.

Light and Colour

Ok, prepare yourself — there's a big plot twist coming up on this page. Hold on to your hats. Wait for it... here we go:

White Light is Not Just a Single Colour

1) White light is actually a mixture of colours.
2) This shows up when white light hits a prism or a rain drop. It gets split up into a full rainbow of colours.
3) The splitting up of light is called dispersal.
4) The proper name for this rainbow effect is a spectrum.

Dispersal of White Light Gives a Spectrum

1) Learn the order that the colours come out in:

    Red Orange Yellow Green Blue Indigo Violet

2) Remember it with this rhyme:

    Richard Of York Gave Battle In Velcro®

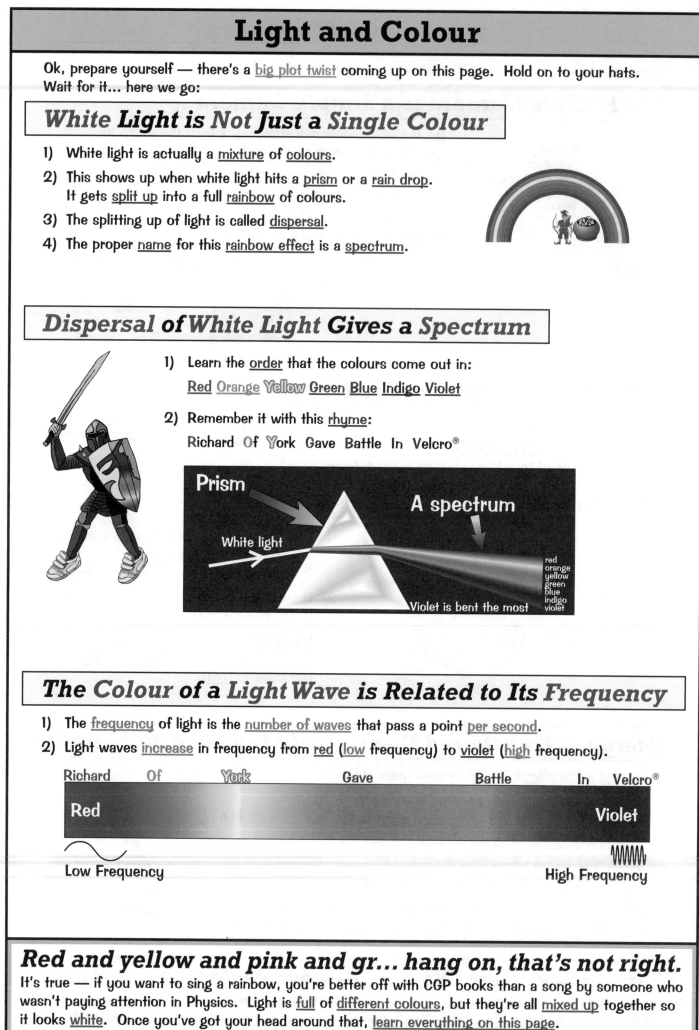

Prism

A spectrum

White light

Violet is bent the most

red
orange
yellow
green
blue
indigo
violet

The Colour of a Light Wave is Related to Its Frequency

1) The frequency of light is the number of waves that pass a point per second.
2) Light waves increase in frequency from red (low frequency) to violet (high frequency).

Richard   Of   York   Gave   Battle   In   Velcro®

Red   Violet

Low Frequency   High Frequency

Red and yellow and pink and gr... hang on, that's not right.

It's true — if you want to sing a rainbow, you're better off with CGP books than a song by someone who wasn't paying attention in Physics. Light is full of different colours, but they're all mixed up together so it looks white. Once you've got your head around that, learn everything on this page.

Absorption and Reflection of Colour

The different colours that make up white light can be <u>reflected</u> and <u>absorbed</u>. Read on...

Coloured *Filters* Only let *Their Colour* Through

1) A <u>filter</u> only allows one <u>particular colour</u> of light to <u>go through it</u>.

2) <u>All other colours</u> are <u>ABSORBED</u> by the filter — so they <u>don't get through</u>.

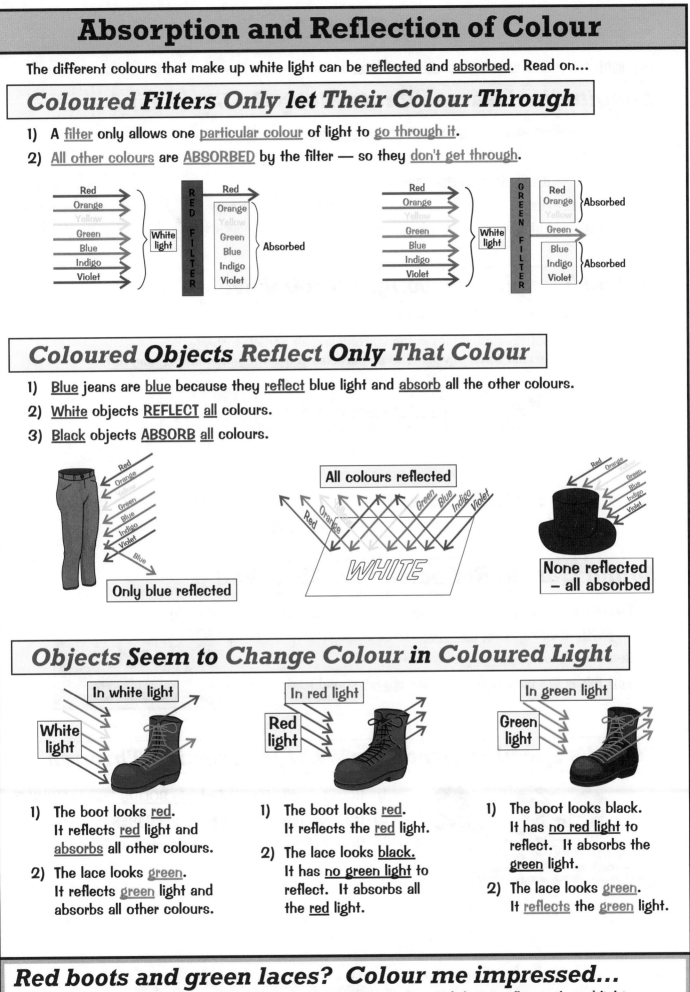

Coloured *Objects* Reflect *Only That Colour*

1) <u>Blue</u> jeans are <u>blue</u> because they <u>reflect</u> blue light and <u>absorb</u> all the other colours.

2) <u>White</u> objects <u>REFLECT</u> <u>all</u> colours.

3) <u>Black</u> objects <u>ABSORB</u> <u>all</u> colours.

Only blue reflected

All colours reflected

WHITE

None reflected – all absorbed

Objects *Seem to Change Colour* in Coloured Light

In white light

White light

In red light

Red light

In green light

Green light

1) The boot looks <u>red</u>.
It reflects <u>red</u> light and <u>absorbs</u> all other colours.

2) The lace looks <u>green</u>.
It reflects <u>green</u> light and absorbs all other colours.

1) The boot looks <u>red</u>.
It reflects the <u>red</u> light.

2) The lace looks <u>black.</u>
It has <u>no green light</u> to reflect. It absorbs all the <u>red</u> light.

1) The boot looks black.
It has <u>no red light</u> to reflect. It absorbs the <u>green</u> light.

2) The lace looks <u>green</u>.
It <u>reflects</u> the <u>green</u> light.

Red boots and green laces? Colour me impressed...

You need to learn this stuff. Luckily it's pretty straightforward — <u>red</u> things reflect only <u>red</u> light, <u>blue</u> things reflect only <u>blue</u> light, and so on. Nevertheless, it's time to cover, scribble, learn...

Sound

Like light, sound is a wave. It's a different type of wave to light though.

Longitudinal Waves Vibrate Along the Same Line

1) Longitudinal waves have vibrations in the same direction as the wave.

2) This means the vibrations are also in the direction of energy transfer.

3) Examples of longitudinal waves include:

   • Sound waves.

   • A slinky spring when you push the end.

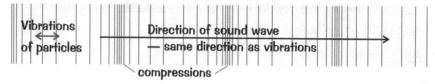

Compressions

Vibrations in same direction as wave is travelling

Sound Travels as a Longitudinal Wave

1) Sound waves are caused by vibrating objects.

2) When an object vibrates, it makes the surrounding air particles vibrate.
   This is how a sound wave travels. The air particles squash together and spread out.

Vibrations of particles

Direction of sound wave — same direction as vibrations

compressions

Compressions are areas of squashed up particles.

3) Sound waves can't travel without particles. So sound needs a medium, like air, to travel.

4) Sound can't travel in space, because it's mostly a vacuum (there are no particles).

Sound Can be Reflected and Absorbed

1) Sound can be reflected and refracted — just like light (see pages 85-86).

2) An ECHO is sound being reflected from a surface.

3) Sound can also be absorbed.

4) Soft things like carpets and curtains absorb sound easily.

Sound's Speed Depends On What It's Passing Through

1) Sound usually travels faster in SOLIDS than in LIQUIDS.

2) It travels faster in liquids than in GASES.

3) Sound always travels much slower than light.

SLOW Air Water Solids FAST
Speed of Sound (e.g. wood)

Learn all this before tea time — sounds like a plan...

Yep. That's another sort of wave for you to get your head around. Sound waves are caused by vibrations — if you've ever put a hand on a bass speaker you'll have 'felt' a sound wave being made.

Hearing

Ear you go then — a whole page on hearing. So pull out your headphones and get learning...

Sound Waves Make Your Ear Drum Vibrate

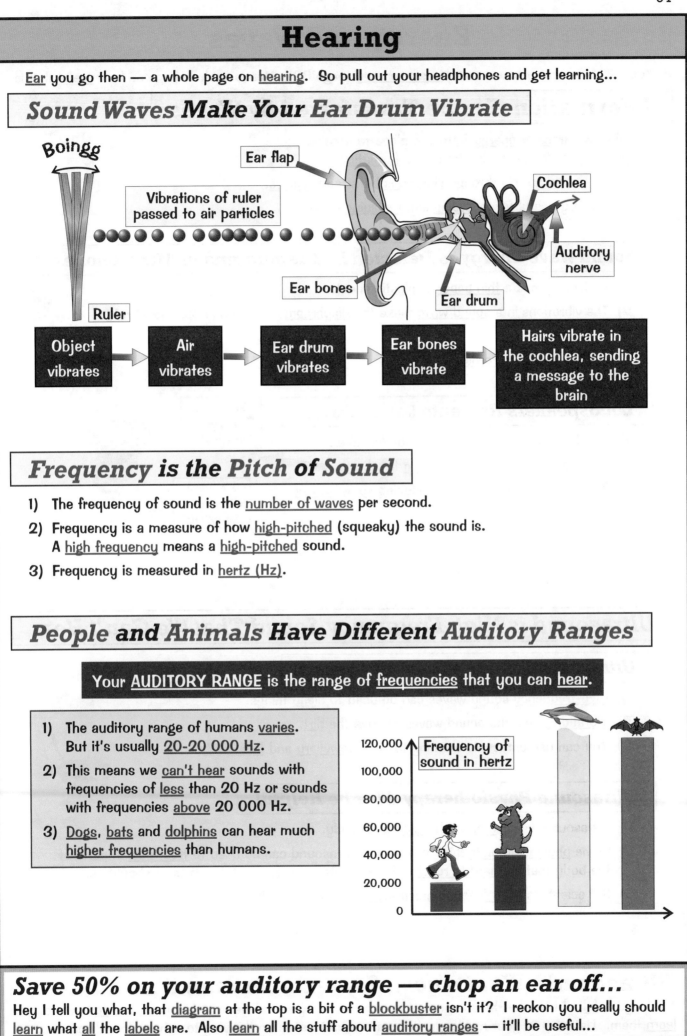

Boingg

Ruler

Vibrations of ruler passed to air particles

Ear flap

Cochlea

Auditory nerve

Ear bones

Ear drum

| Object vibrates | → | Air vibrates | → | Ear drum vibrates | → | Ear bones vibrate | → | Hairs vibrate in the cochlea, sending a message to the brain |

Frequency is the Pitch of Sound

1) The frequency of sound is the <u>number of waves</u> per second.

2) Frequency is a measure of how <u>high-pitched</u> (squeaky) the sound is. A <u>high frequency</u> means a <u>high-pitched</u> sound.

3) Frequency is measured in <u>hertz (Hz)</u>.

People and Animals Have Different Auditory Ranges

Your <u>AUDITORY RANGE</u> is the range of <u>frequencies</u> that you can <u>hear</u>.

1) The auditory range of humans <u>varies</u>. But it's usually <u>20-20 000 Hz</u>.

2) This means we <u>can't hear</u> sounds with frequencies of <u>less</u> than 20 Hz or sounds with frequencies <u>above</u> 20 000 Hz.

3) <u>Dogs</u>, <u>bats</u> and <u>dolphins</u> can hear much <u>higher frequencies</u> than humans.

Frequency of sound in hertz

120,000
100,000
80,000
60,000
40,000
20,000
0

Save 50% on your auditory range — chop an ear off...

Hey I tell you what, that <u>diagram</u> at the top is a bit of a <u>blockbuster</u> isn't it? I reckon you really should <u>learn</u> what <u>all</u> the <u>labels</u> are. Also <u>learn</u> all the stuff about <u>auditory ranges</u> — it'll be useful...

Energy and Waves

You might remember how waves transfer energy (page 83). Here's a whole page on how useful that is.

Information *Can be Transferred by Waves*

1) All waves transfer energy from one place to another.
   In doing so, they can also transfer information.

2) Sound waves do this through vibrations between particles.

3) This is very useful for recording and replaying sounds.

Sound Wave Energy *is Detected by Diaphragms in Microphones*

1) A diaphragm is a thin paper or plastic sheet.

2) The vibrations in a sound wave make the diaphragm inside a microphone vibrate.

3) The microphone converts the vibrations to electric signals.

4) Another device records the signals.

Loudspeakers Recreate Sound Waves

1) An electrical signal is fed into a loudspeaker.

2) This signal causes the diaphragm to vibrate.

3) This makes the air vibrate, producing sound waves.

It's a bit like a microphone in reverse.

Diaphragm makes air vibrate

Electrical pulses cause vibrations

Ultrasound *is High Frequency Sound That We Can't Hear*

Ultrasonic Cleaning *Uses Ultrasound*

1) High-frequency sound waves can be used to clean things.

2) Vibrations from the sound waves remove the dirt.

3) You can use ultrasonic cleaning to clean jewellery and false teeth.

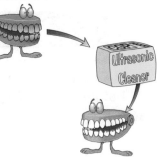

Ultrasound Physiotherapy *May be Helpful*

1) Ultrasound waves can reach inside your body.

2) Some physiotherapists think this means ultrasound can be used to treat pain in parts of the body that are hard to reach. For example, muscles deep inside your shoulders.

3) But scientists haven't found much evidence that this actually works.

Ultrasound? What's next, SuperDuperSound?

This page is full of uses for the energy that is transferred by sound waves. And yes, you do have to learn them. Listening to your music player create sound waves on the way to school doesn't count...

Section Summary

Section 10 tells you everything you need to know about waves. To really ace this section, you'll need to have a go at all these questions. If there are any answers you don't know, don't panic — just look them up. All the answers are somewhere in this section. Once you've done all the questions once, have another go. And then another. Keep going and, before you know it, you'll be a total pro at waves.

1) What type of wave is a water wave?
2) What do water waves look like? Sketch out a diagram and label it.
3) Give three things water waves and light waves have in common.
4) Give one big difference between water waves and light waves.
5) What speed does light travel at in a vacuum?
6) Why do rough surfaces look dull?
7) What is the law of reflection?
8) What does the word "medium" mean?
9) What is refraction?
10) What happens when light goes from a less dense medium to a <u>more</u> dense medium?
11) What happens when light goes from a more dense medium to a <u>less</u> dense medium?
12) Why does light bend when it enters a glass block?
13) Sketch a diagram of a pinhole camera.
14) Why is the image seen through a pinhole camera upside down and crossed over?
15) What does a convex lens look like? What does it do?
16) Which two parts of the eye help you focus on an object?
17) How do digital cameras form images?
18) How could you show that white light is not just one colour?
19) What is the rhyme for remembering the order of colours in a spectrum?
20) What colour of light has the highest frequency?
21) What colour of light will a red filter let through?
22) Why does something blue look blue in white light?
23) What happens to all the colours in white light when they hit a black object?
24) What colour would green laces look in red light? Why?
25) What type of wave are sound waves?
26) In which direction are the vibrations?
27) What does sound need to travel from one place to another?
28) What is an echo?
29) Does sound usually travel faster in solids, liquids or gases?
30) Explain how a ruler being flicked can be heard.
31) What does the frequency of a sound mean?
32) What does auditory range mean?
33) What is the auditory range of humans?
34) How do microphones work?
35) What is ultrasound?
36) What can it be used for?

Electrical Circuits

First up in this section, some <u>electricity basics</u>...

Electricity Flows Through Circuits

1) This is an <u>electric circuit</u>.
2) It has a <u>power supply</u>, <u>wires</u>, and a <u>component</u> (the bulb).
3) Electric current <u>flows</u> from the power supply in a <u>loop</u> around the circuit and back to the power supply.

Electric Current is the Flow of Charge

1) <u>Current</u> is the <u>flow</u> of <u>charge</u> around a circuit. The moving charges are <u>negative electrons</u>.
2) Current can only flow if a circuit is <u>complete</u> with <u>no gaps</u>.
3) <u>CURRENT ISN'T USED UP</u> as it flows through a circuit. The <u>total current</u> is always the <u>same</u>.

> Current is a bit like water flowing...
> 1) The pump drives the <u>water along</u> like a power supply.
> 2) The water <u>isn't used up</u> — <u>all</u> of it arrives back at the <u>pump</u>.

Potential Difference Pushes the Current Around

1) In a circuit, the <u>battery</u> provides the <u>driving force</u> that <u>pushes</u> the charge round the circuit.
2) This driving force is called the <u>potential difference</u>.
3) If you <u>increase</u> the potential difference <u>more current</u> will flow.

Potential difference is sometimes called voltage.

Resistance is How Easily Electricity Can Flow

1) <u>Resistance slows down</u> the flow of current. It's measured in <u>ohms</u> (Ω).
2) The <u>resistance</u> of a component is equal to the <u>POTENTIAL DIFFERENCE divided by</u> the <u>CURRENT</u>.
3) <u>Conductors</u> are materials that allow electricity to pass through them <u>easily</u> — such as <u>metals</u>.
4) <u>Insulators</u> are materials that <u>don't</u> allow electricity to pass through them easily — such as <u>wood</u>.
5) The <u>lower the resistance</u> of a component, the <u>better</u> it is at <u>conducting electricity</u>.

1) The <u>metal strip</u> has a resistance of <u>0.001 Ω</u> — it's a good <u>conductor</u>.
2) But the <u>wooden block</u> has a <u>very high</u> resistance — it's an <u>insulator</u>.

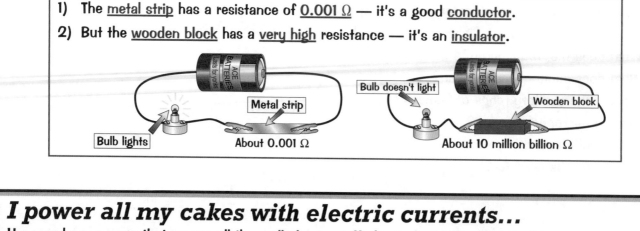

I power all my cakes with electric currents...

Here we have a page that covers all the <u>really basic stuff</u> about electricity. Get it all <u>learnt</u>.

Measuring Current and Potential Difference

Sadly you don't just need to know what <u>current</u> and <u>potential difference</u> are — you need to be able to <u>measure</u> them too. Handily, some clever chaps have made machines to do just that...

Ammeters Measure Current

1) <u>Ammeters</u> measure electric <u>current</u>.
It's measured in <u>amperes</u>, **A**.

2) Remember — current <u>doesn't</u> get used up in a circuit.
So the current through the ammeter and the bulb are the <u>same</u>.

3) You can connect the ammeter <u>anywhere</u> in the circuit.

Voltmeters Measure Potential Difference

1) <u>Voltmeters</u> measure <u>potential difference</u> in <u>volts</u>, **V**.

2) You measure the potential difference <u>between two points</u> in the circuit. For example, <u>either side</u> of a bulb.

Batteries and Bulbs Have Potential Difference Ratings

1) A <u>battery</u> potential difference rating tells you the <u>potential difference</u> it will <u>supply</u>.

2) A <u>bulb rating</u> tells you the <u>maximum</u> potential difference that you can <u>safely</u> put across it.

Battery rating

Bulb rating

1.5 V   2.5 V

Circuit Diagrams Represent Real Circuits

1) Circuit <u>diagrams</u> are <u>simplified drawings</u> of real circuits.

2) Here are the circuit <u>symbols</u> you <u>need to know</u>:

A cell = ⊣|⊢
(a single energy source)

> In everyday life we call a cell a battery.

A battery = ⊣|⊦|⊢
(a battery is two or more cells put together)

A switch:
– open = ⊸⁄⊸   – closed = ⊸⊸

A bulb = –⊗–

A voltmeter = –(V)–

An ammeter = –(A)–

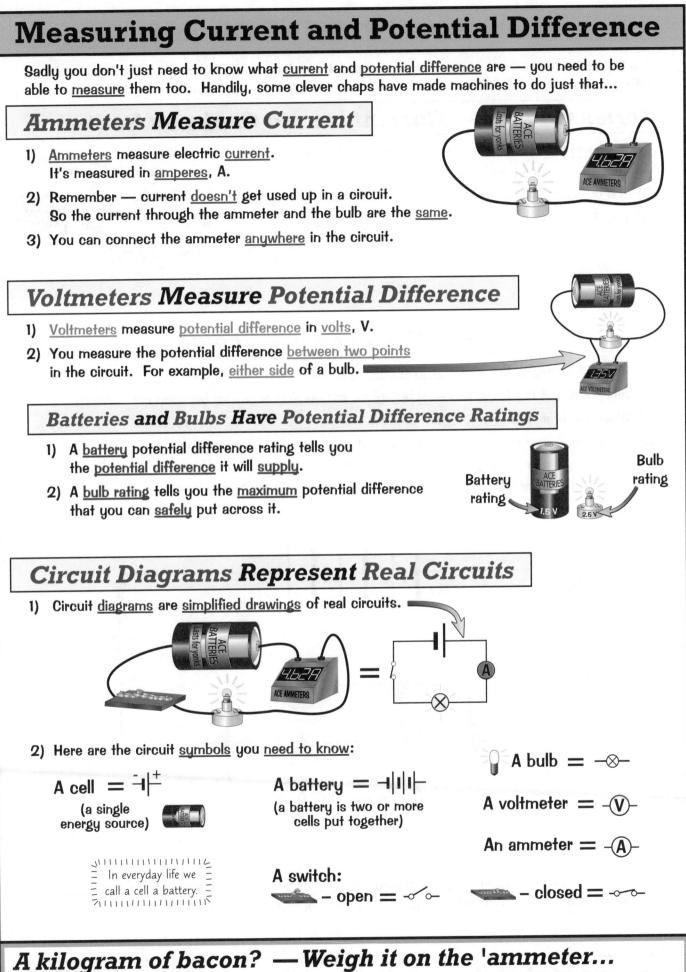

A kilogram of bacon? — Weigh it on the 'ammeter...

You'll need to get your ruler out to draw a nice, neat <u>circuit diagram</u>. They don't look much like the real circuits they show — but they do make it <u>easier</u> to see how everything is connected up. The reason I like them is that symbols are so much easier to draw than the actual components.

Series and Parallel Circuits

The big difference between <u>series</u> and <u>parallel</u> circuits is that in parallel circuits, current can take <u>different routes</u> around the circuit. And the charges don't even need a map or a GPS to do it...

*Series Circuits — Current **has** No Choice **of** Route*

1) In a <u>series</u> circuit the current has <u>no choice</u> of <u>route</u>. There is only <u>one way</u> it can go around the <u>circuit</u>.

2) The current gives up <u>some</u> of its <u>energy</u> to <u>each</u> of the <u>bulbs</u>.

3) But the <u>current stays the same</u> all the way around.

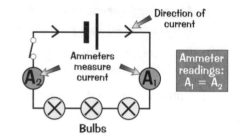

*Parallel Circuits — Current **has** a Choice*

1) In <u>parallel circuits</u>, there's <u>more than one route</u> that current could take.

2) When a circuit divides into several branches, the <u>current</u> is <u>divided</u> between each branch.

3) When the branches <u>join up</u> again, so does the <u>current</u>.

4) The current <u>after</u> the join is just the current of each branch <u>added together</u>.

5) Don't forget that current <u>isn't used up</u>. So <u>after</u> the branches join, the current is the <u>same</u> as it was <u>before</u> it split.

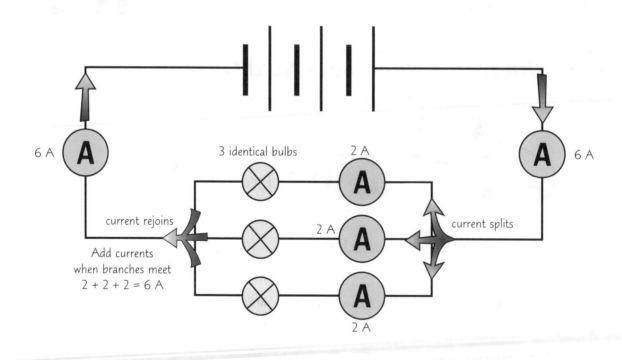

6) Parallel circuits are <u>great</u> — you can turn off <u>parts</u> of them <u>without</u> turning off the <u>whole circuit</u>. This means you can turn off <u>one</u> light without having to turn off <u>everything</u> in your house. <u>Useful</u>.

A series of circuits — well, there are two on this page...

Circuits cause people a lot of grief, that's for sure. The worst thing about them is that you can't actually <u>see</u> the current flowing, so it's very difficult to appreciate what's <u>going on</u>. Tough stuff.

Static Electricity

Right, that's enough on charges flowing about the place. Now let's look at <u>static</u> charges...

Charges *Can* Build Up *if Objects* are Rubbed Together

1) <u>Atoms</u> (see page 34) contain <u>positive</u> and <u>negative charges</u>.

2) The <u>negative</u> charges are called <u>electrons</u>. <u>Electrons</u> can <u>move</u>, but <u>positive charges can't</u>.

3) When two insulating objects (see page 94) are <u>rubbed</u> together, the <u>electrons</u> are <u>scraped off</u> one object and <u>left</u> on the other.

4) This gives both objects a <u>static charge</u>:

> • The object that <u>gains electrons</u> becomes <u>negatively</u> charged.
> • The object that <u>loses electrons</u> is left with an <u>equal</u> but <u>positive</u> charge.

A static charge doesn't move.

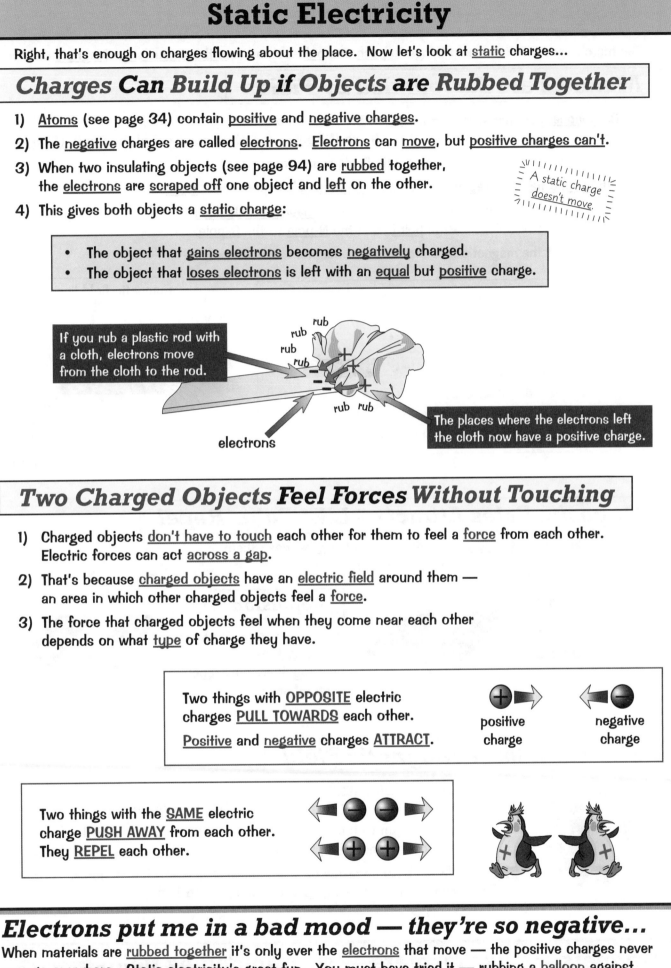

If you rub a plastic rod with a cloth, electrons move from the cloth to the rod.

electrons

The places where the electrons left the cloth now have a positive charge.

Two Charged Objects *Feel Forces* Without Touching

1) Charged objects <u>don't have to touch</u> each other for them to feel a <u>force</u> from each other. Electric forces can act <u>across a gap</u>.

2) That's because <u>charged objects</u> have an <u>electric field</u> around them — an area in which other charged objects feel a <u>force</u>.

3) The force that charged objects feel when they come near each other depends on what <u>type</u> of charge they have.

Two things with <u>OPPOSITE</u> electric charges <u>PULL TOWARDS</u> each other.
<u>Positive</u> and <u>negative</u> charges <u>ATTRACT</u>.

positive charge
negative charge

Two things with the <u>SAME</u> electric charge <u>PUSH AWAY</u> from each other.
They <u>REPEL</u> each other.

Electrons put me in a bad mood — they're so negative...

When materials are <u>rubbed together</u> it's only ever the <u>electrons</u> that move — the positive charges never ever go anywhere. Static electricity's great fun. You must have tried it — rubbing a <u>balloon</u> against your <u>head</u> and getting your <u>hair</u> to stick up like a crazy scientist's. Your hair sticks up like that because each of your hairs has the <u>same type of charge</u>, so they <u>repel</u> each other. Neat.

Magnets

Electric charges aren't the only things to <u>push</u> and <u>pull</u> each other <u>without touching</u>. <u>Magnets</u> do it too.

Magnets are Surrounded by Fields

1) <u>Bar magnets</u> are <u>magnets</u> that are in the shape of a bar.
One end of a bar magnet is called the <u>North (N) pole</u>. The other end's called the <u>South (S) pole</u>.

2) All bar magnets have <u>invisible magnetic fields</u> round them.

3) A <u>magnetic field</u> is a <u>region</u> where <u>magnetic materials</u> experience a <u>force</u>.

4) You can draw a magnetic field using lines called <u>magnetic field lines</u>.
The magnetic field lines always <u>point</u> from the <u>N-pole</u> to the <u>S-pole</u>.

5) This is what the magnetic field around a bar magnet looks like:

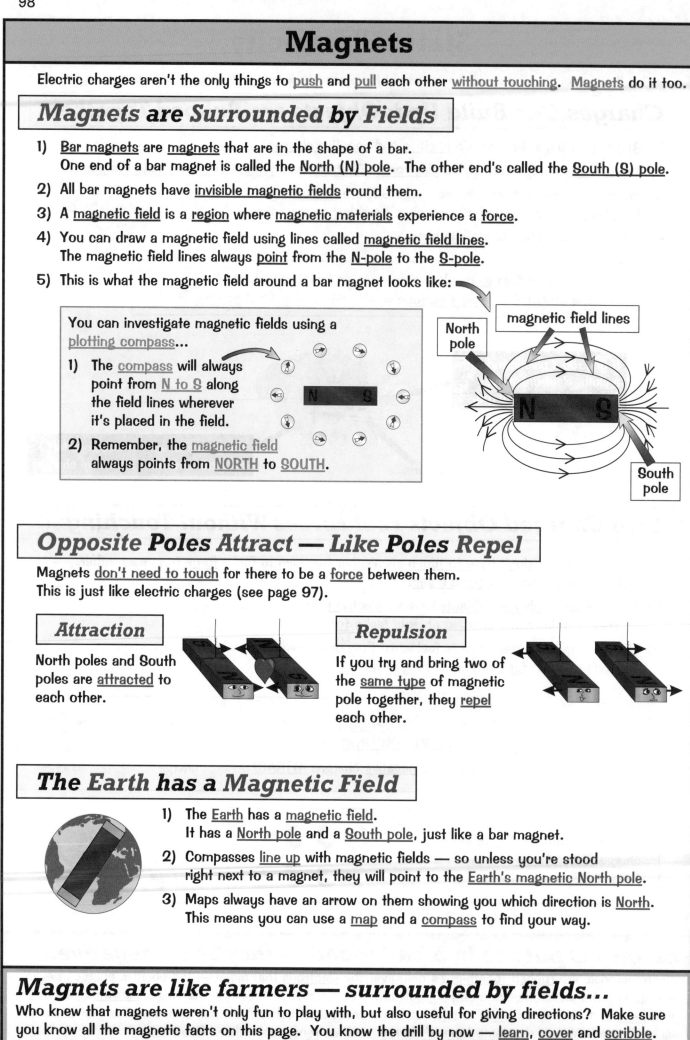

You can investigate magnetic fields using a <u>plotting compass</u>...

1) The <u>compass</u> will always point from <u>N to S</u> along the field lines wherever it's placed in the field.

2) Remember, the <u>magnetic field</u> always points from <u>NORTH</u> to <u>SOUTH</u>.

North pole

magnetic field lines

South pole

Opposite Poles Attract — Like Poles Repel

Magnets <u>don't need to touch</u> for there to be a <u>force</u> between them.
This is just like electric charges (see page 97).

Attraction

North poles and South poles are <u>attracted</u> to each other.

Repulsion

If you try and bring two of the <u>same type</u> of magnetic pole together, they <u>repel</u> each other.

The Earth has a Magnetic Field

1) The <u>Earth</u> has a <u>magnetic field</u>.
It has a <u>North pole</u> and a <u>South pole</u>, just like a bar magnet.

2) Compasses <u>line up</u> with magnetic fields — so unless you're stood right next to a magnet, they will point to the <u>Earth's magnetic North pole</u>.

3) Maps always have an arrow on them showing you which direction is <u>North</u>.
This means you can use a <u>map</u> and a <u>compass</u> to find your way.

Magnets are like farmers — surrounded by fields...

Who knew that magnets weren't only fun to play with, but also useful for giving directions? Make sure you know all the magnetic facts on this page. You know the drill by now — <u>learn</u>, <u>cover</u> and <u>scribble</u>.

Electromagnets

Bar magnets stay magnetic all the time. Electromagnets are magnets which you can turn on and off.

Electric Current in Wires Causes a Magnetic Field

1) An electric current going through a wire causes a magnetic field around the wire.

2) A long coil of wire with a current flowing through it has a magnetic field just like a bar magnet's.

3) Magnets made from a current-carrying wire are called ELECTROMAGNETS.

4) They're usually made from a coil of wire wrapped around a soft iron core.

5) Because you can turn the current on and off, the magnetic field can be turned on and off.

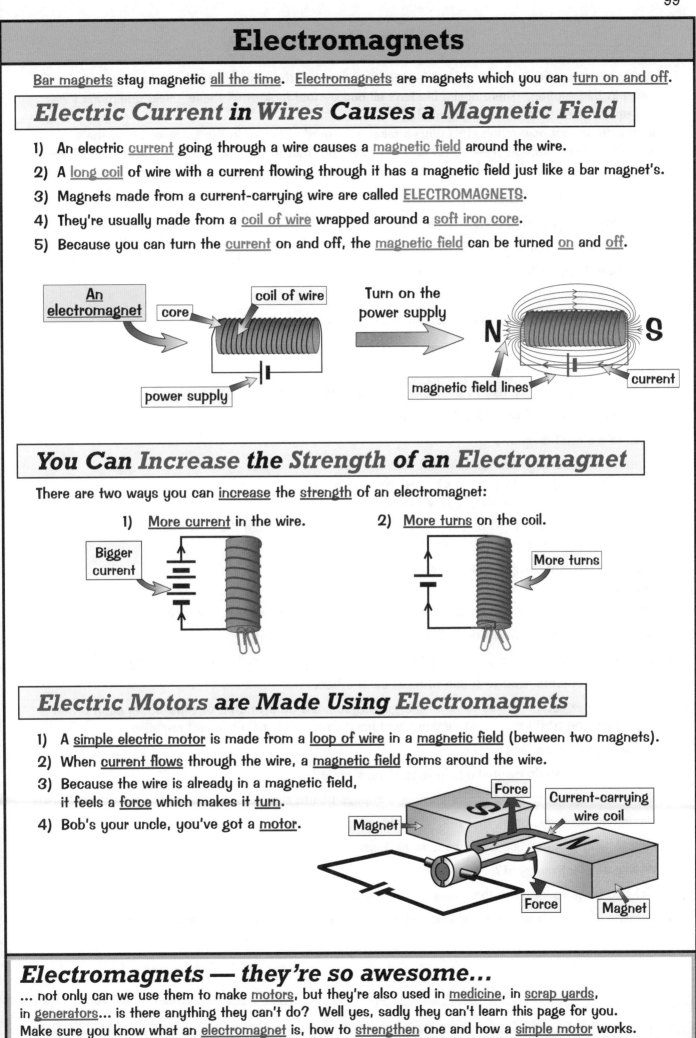

You Can Increase the Strength of an Electromagnet

There are two ways you can increase the strength of an electromagnet:

1) More current in the wire.

2) More turns on the coil.

Electric Motors are Made Using Electromagnets

1) A simple electric motor is made from a loop of wire in a magnetic field (between two magnets).

2) When current flows through the wire, a magnetic field forms around the wire.

3) Because the wire is already in a magnetic field, it feels a force which makes it turn.

4) Bob's your uncle, you've got a motor.

Electromagnets — they're so awesome...

... not only can we use them to make motors, but they're also used in medicine, in scrap yards, in generators... is there anything they can't do? Well yes, sadly they can't learn this page for you. Make sure you know what an electromagnet is, how to strengthen one and how a simple motor works.

Section Summary

Electricity and Magnetism — it's no holiday, that's for sure. There are certainly quite a few nasty bits and bobs in this section. There again, life isn't all bad — just look at all these lovely questions I've cooked up for your delight and enjoyment.

These questions test how much stuff you've taken on board. They're in the same order as the stuff appears throughout Section 11 — so for any you can't do, just look back, find the answer, and then learn it good and proper for next time.

1) Current is the flow of what?

2) Can current flow in an incomplete circuit?

3) What job does a battery do in a circuit?

4) What is potential difference?

5) What is resistance?

6) What is the difference between a conductor and an insulator?

7) What instrument do we use to measure current?

8) What are the units of current?

9) What instrument do we use to measure potential difference?

10) What are the units of potential difference?

11) What is a circuit diagram?

12) Sketch the circuit symbol for all of these:
   a) a bulb   b) a battery   c) a switch (open)
   d) a cell   e) an ammeter   f) a voltmeter.

13)*A series circuit contains 3 bulbs. A current of 3 A flows through the first bulb. What current flows through the third bulb?

14) In parallel circuits current has a choice of what?

15) *True or false?
    Adding the current through each branch of a parallel circuit gives you the total current.

16) Which type of circuit allows part of the circuit to be switched off?

17) Explain how a cloth and a plastic rod both become charged when they're rubbed together.

18) Do charged objects need to touch to repel each other?

19) State whether each of these pairs of charged objects will be attracted or repelled by each other.
    a) positive and positive   b) negative and positive   c) negative and negative

20) What is a magnetic field?

21) In which direction do magnetic field lines always point?

22) Sketch a diagram showing how a plotting compass points around a bar magnet.

23) Name two magnetic poles that will: a) attract each other   b) repel each other.

24) Explain why you can use a compass to navigate.

25) What's an electromagnet?

26) List two ways to increase the strength of an electromagnet.

27) The wire in a simple motor feels a force when a current flows through it. Why is this?

* Answers on page 108.

Gravity

It's not <u>magic</u> that keeps your feet on the ground, it's <u>gravity</u>. As Sandra Bullock will tell you.

Gravity is a Force that Attracts All Masses

1) Anything with <u>mass</u> will <u>attract</u> anything else with mass.

2) The force of attraction between two objects is called <u>GRAVITY</u>.

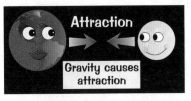

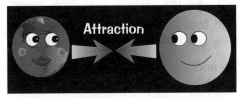

The <u>Earth</u> and the <u>Moon</u> are <u>attracted by gravity</u>.

The <u>Earth</u> and the <u>Sun</u> are attracted by an even <u>bigger force</u> of <u>gravity</u>.

3) Objects with a <u>bigger mass</u> attract each other with a <u>stronger force</u>.

4) <u>Gravitational field strength</u> (g) is how <u>strong</u> gravity is. It's <u>different</u> on <u>different planets</u> and <u>stars</u>.

Learn this: ➡ On <u>Earth</u> g = <u>10 N/kg</u>.

On <u>Mars</u> g = <u>3.7 N/kg</u>.
Gravity is <u>weaker on Mars</u> than on Earth.

Learn the Difference Between Mass and Weight

MASS	WEIGHT
• <u>Mass</u> is the <u>amount of 'stuff'</u> in an object. • Mass is <u>NOT</u> a <u>force</u>.	• <u>Weight</u> is a <u>FORCE</u>. • It is caused by the <u>pull</u> of <u>GRAVITY</u>.
• The mass of an object <u>never changes</u>, no matter where it is in the Universe.	• The weight of an object is <u>different</u> on <u>different planets</u> and <u>stars</u>.
• Mass is measured in <u>kilograms</u> (<u>kg</u>) using a <u>mass balance</u>.	• Weight is measured in <u>newtons</u> (<u>N</u>) using a <u>spring balance</u> or a <u>newton meter</u>.

Now Learn this Formula...

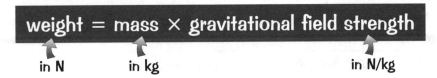

weight = mass × gravitational field strength

in N   in kg   in N/kg

<u>Example 1</u>: An object has a mass of 5 kg. What is its weight on Earth, in newtons?

<u>Answer</u>: On Earth g = <u>10 N/kg</u>. So the weight of the 5 kg object = 5 × 10 = <u>50 N</u>.

<u>Example 2</u>: An object has a mass of 5 kg. What is its weight on Mars, in newtons?

<u>Answer</u>: On Mars g = <u>3.7 N/kg</u>. So the weight of the 5 kg object = 5 × 3.7 = <u>18.5 N</u>.

In these examples, the object <u>always has a MASS of 5 kg</u>.
But the <u>WEIGHT</u> of the object is <u>different</u> on <u>Earth</u> and on <u>Mars</u>.

The Sun and Stars

Ahh. This is going to be a <u>nice page</u>, I can tell. Look at all those <u>lovely pictures</u> for a start.

The Earth Moves Round the Sun

1) The <u>Sun</u> is a <u>star</u>.
2) The Earth is a <u>planet</u>.
3) The <u>Earth</u> moves round the Sun in a <u>rough circle</u>. This circle is called an <u>ORBIT</u>.
4) The <u>Sun</u> and other <u>stars</u> give out <u>light</u>. Planets <u>don't</u>.

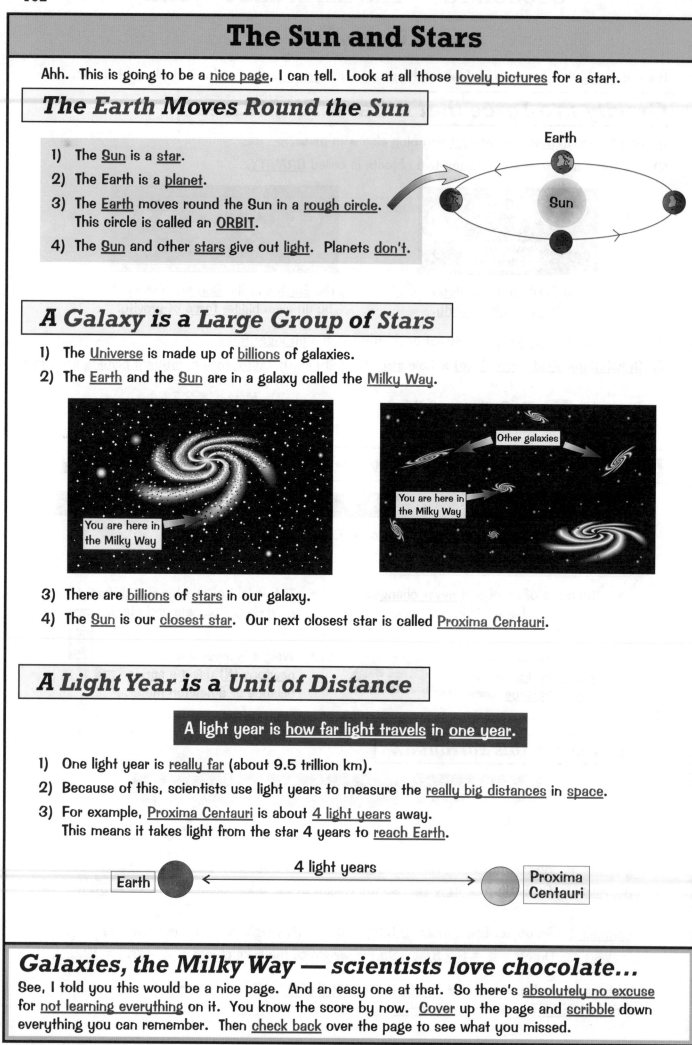

Earth

Sun

A Galaxy is a Large Group of Stars

1) The <u>Universe</u> is made up of <u>billions</u> of galaxies.
2) The <u>Earth</u> and the <u>Sun</u> are in a galaxy called the <u>Milky Way</u>.

You are here in the Milky Way

Other galaxies

You are here in the Milky Way

3) There are <u>billions</u> of <u>stars</u> in our galaxy.
4) The <u>Sun</u> is our <u>closest star</u>. Our next closest star is called <u>Proxima Centauri</u>.

A Light Year is a Unit of Distance

A light year is <u>how far light travels</u> in <u>one year</u>.

1) One light year is <u>really far</u> (about 9.5 trillion km).
2) Because of this, scientists use light years to measure the <u>really big distances</u> in <u>space</u>.
3) For example, <u>Proxima Centauri</u> is about <u>4 light years</u> away. This means it takes light from the star 4 years to <u>reach Earth</u>.

4 light years

Earth

Proxima Centauri

Galaxies, the Milky Way — scientists love chocolate...

See, I told you this would be a nice page. And an easy one at that. So there's <u>absolutely no excuse</u> for <u>not learning everything</u> on it. You know the score by now. <u>Cover</u> up the page and <u>scribble</u> down everything you can remember. Then <u>check back</u> over the page to see what you missed.

Day and Night and the Four Seasons

There's a fair bit to <u>learn</u> on this page. So let's get cracking...

The Rotation of The Earth Causes Day and Night

1) The Earth <u>rotates</u> (turns) about its <u>axis</u> —
   an imaginary line running through its centre,
   from the North Pole to the South Pole.

2) A <u>globe</u> does the same thing.

3) As the Earth rotates, any place on its surface
   will <u>sometimes face the Sun</u>.
   At other times it will <u>face away</u>.

4) When a place <u>faces the Sun</u> it gets <u>light</u> —
   so it's <u>day time</u>.

5) When a place <u>faces away</u> from the Sun it gets <u>no</u> light — so it's <u>night time</u>.

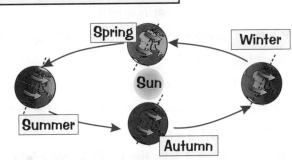

The Seasons are Caused by the Earth's Tilt

1) The Earth takes <u>one year</u> to <u>orbit</u> the Sun <u>once</u>.

2) Each year has <u>four seasons</u> —
   <u>SPRING</u>, <u>SUMMER</u>, <u>AUTUMN</u> and <u>WINTER</u>.

3) The seasons are caused by the <u>tilt</u> (angle)
   of the <u>Earth's axis</u>.

Summer

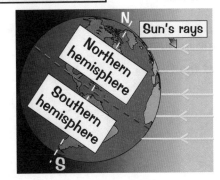

1) When it's summer in the UK, the <u>northern hemisphere</u>
   (top half of the Earth) is tilted <u>towards</u> the Sun.

2) This means the <u>sunlight</u> we get is <u>stronger</u>
   and we get <u>more hours</u> of it.

3) This gives us <u>longer</u>, <u>warmer days</u>
   — and we have <u>summer</u>.

Winter

1) When it's winter in the UK, the northern
   hemisphere is tilted <u>away</u> from the Sun.

2) This means the <u>sunlight</u> we get is
   <u>weaker</u> and we get <u>fewer hours</u> of it.

3) This gives us <u>shorter</u>, <u>colder days</u>
   — and we have <u>winter</u>.

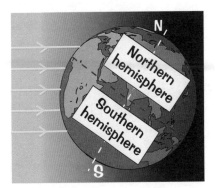

When it's <u>summer</u> in the <u>northern hemisphere</u>, it's <u>winter</u> in the <u>southern hemisphere</u>.

Phew — I feel quite giddy now...
Make sure you know why the tilt of the Earth's axis gives us <u>warm summers</u> and <u>cold winters</u>.

Section Summary

Section 12 only has three pages of information — and it deals with the whole Universe. It's amazing just how many people go their whole lives and never really know the answers to all those burning questions, like what is gravity? Or why are the days longer in summer than in winter?
Make sure you learn all the burning answers now...

1) What is gravity?

2) Which is stronger:
   a) the force of gravity between the Moon and the Earth?   OR
   b) the force of gravity between the Sun and the Earth?

3) What is gravitational field strength (g)?

4) On Earth, what does 'g' equal?

5) What is mass? Is it a force?

6) What is weight? Is it a force?

7) What unit is mass measured in? What unit is weight measured in?

8)*On Jupiter, g = 25 N/kg. What would a 5 kg object weigh on Jupiter?
   Remember to include the correct unit in your answer.

9) Name the star the Earth orbits.

10) Which of these is given out by stars? a) chocolate,   b) light,   c) cold air.

11) What is a galaxy?

12) What is the name of our galaxy?

13) Apart from the Sun, name one other star in our galaxy.

14) What is a light year?

15) How long does it take the Earth to complete one full rotation on its axis?

16) The UK is facing the Sun. Is it day time or night time there?

17) Australia is facing away from the Sun. Is it day time or night time in Australia?

18) How long does it take the Earth to complete one full orbit around the Sun?

19) How many seasons are there?

20) Why do we get more hours of sunlight in summer?

21) Give two reasons why we get shorter, colder days in winter.

* Answer on page 108.

Index

Index

Index

Index and Answers

Answers to Selected Section Summary Questions

Section Summary 2 — Page 18
6) Daily basic energy requirement = 5.4 × 24 hours × body mass (kg)
= 5.4 × 24 × 54 = 6998.4 kJ

Section Summary 3 — Page 25
23) There would be fewer perch around to eat the water beetles.

Section Summary 5 — Page 44
16) Ca = calcium, Cl = chlorine

Section Summary 6 — Page 55
12) exothermic   **13)** endothermic

Section Summary 8 — Page 74
6) The crane that applies the small force will lift the weight furthest.
15) Energy transferred = power (kW) × time (hours)
= 1.5 kW × 1 hour
= 1.5 kWh
17) Cost = energy transferred in kWh × price per kWh
= 298.2 × 15 = 4473p = £44.73
19) The 300 W device (it has a higher power rating).
25) 50 g (the amount of substance is the same before and after).

Section Summary 9 — Page 82
3) speed = distance ÷ time = 5 ÷ 2 = 2.5 m/s.
25) 200 ÷ 2 = 100 N/m^2 or 100 Pa
27) The boat will sink.

Section Summary 11 — Page 100
13) 3 A   **15)** True

Section Summary 12 — page 104
8) weight = mass × gravitational field strength (g)
weight = 5 × 25 = 125 N